Ecological Stylistics

Daniela Francesca Virdis

Ecological Stylistics

Ecostylistic Approaches to Discourses of Nature, the Environment and Sustainability

palgrave
macmillan

Daniela Francesca Virdis
Department of Humanities, Languages
and Heritage
University of Cagliari
Cagliari, Italy

ISBN 978-3-031-10657-6 ISBN 978-3-031-10658-3 (eBook)
https://doi.org/10.1007/978-3-031-10658-3

Cover credit: Carmine Sau & Francesco Virdis

This Palgrave Macmillan imprint is published by the registered company Springer Nature Switzerland AG
The registered company address is: Gewerbestrasse 11, 6330 Cham, Switzerland

To Carmine,
for teaching me the difference
between Quercus ilex *and* Quercus suber.

Acknowledgements

While I was researching and writing this book, various colleagues and friends were extremely helpful and generously provided me with their advice and support.

The contributors to the volumes I recently co-edited (*Language in Place: Stylistic Perspectives on Landscape, Place and Environment* [2021] and *The Stylistics of Landscapes, the Landscapes of Stylistics* [2017]) and those to the Special Issue I am editing (*Journal of World Languages*, due in 2023) inspired this book, its contents, theories and approaches with their ground-breaking research in the field of stylistics, and with their analyses of the environment and related topics in literary and non-literary texts.

A lot of food for thought was offered by the innovative work and ideas of Ernestine Lahey, Elisabetta Zurru and all the members of LAND-SIG, a Poetics And Linguistics Association (PALA) Special Interest Group on the stylistics of landscape, place and environment, coordinated by Ernestine, Elisabetta and myself.

Several members of PALA and of the International Association of Literary Semantics (IALS) brought to my attention their research on place and landscape explored from a linguistic perspective: their work is reviewed here in the theoretical Chapter 2, along with the most recent literature in ecostylistics, and contributed to defining the aims and scope of the area.

Profound and refreshing insights were supplied by the keynote speakers and the participants in the International Conferences on Ecolinguistics ICE-4 (Odense, 2019), organised by Sune Vork Steffensen, and ICE-5 (Liverpool, 2021), organised by Jessica Hampton, Mariana Roccia, Camila Montiel McCann and Madeleine Mancey.

Steve Buckledee and the anonymous reviewer of this book provided me with penetrating comments and solid advice on a previous draft of this work.

Alwin Fill and Michael Toolan wrote extremely generous endorsements of this research and had very kind words about its final results.

Cathy Scott and Uma Vinesh at Palgrave Macmillan and Springer Nature believed in this project and positively contributed to ensuring that it saw the light of day.

For helping me to bring this book to fruition, I owe a debt of deep gratitude to all these people.

Contents

1

Introduction: Towards an Ecological Stylistics

1.1 Background, Main Research Purpose and Contents

This book reflects the cutting edge in ecostylistic approaches to nature, the environment and sustainability as represented in contemporary non-literary discourse: it therefore showcases this emerging and burgeoning field at the forefront of work in both ecolinguistics and stylistics. It is the first book-length study complying with the theoretical framework and methodology of ecostylistics; it is also the first attempt to systematically apply ecostylistics to the examination of present-day discourses about ecological issues and the environmental crisis (for this term, see Sect. 1.3). In short, it provides the first thorough paradigm for the subject area of ecostylistics and an analytical approach to non-literary texts considering ecological questions.[1]

The texts investigated in this book are a part of the contemporary social and cultural environet, or "the aggregate collection of texts, words and voices [about the environment], a network making linkage upon linkage between the environment words" and discourses (Myerson and Rydin 1996, 7). More precisely,

D. F. Virdis, *Ecological Stylistics*,
https://doi.org/10.1007/978-3-031-10658-3_1

> In the environet, texts always interact with each other, indirectly because they address issues together, they use the same words, they address similar audiences, they respond to similar feelings, they use the same arguments, or opposing arguments [...]. The effect of these repeated connections is cumulative, building up support for different viewpoints and pushing certain issues into more prominent positions in the environmental agenda. As arguments accumulate across the environet, the agendas reform. Therefore, any single text has broader impacts than are felt by its readers; each contributes to the aggregate impacts of environmental texts on the culture as a whole. (Myerson and Rydin 1996, 10)

The most relevant words in the environet were called "marker words" or "environment words" by Myerson and Rydin (1996, 6 and 37), since they reveal an ecological agenda, carry multiple viewpoints and voices, and mark the topics, arguments and challenges composing the environet. To be more precise,

> Each word marks a cluster of arguments, a topic; and each topic would be a different way to enter the net, to enter from another angle. It is always the same net, but it is viewed differently from whichever topic is chosen. In addition, the arguments are processes. Each topic is recreated by every argument, and so the relations between topics are continuously in motion. The environmental topics are a system, but a changing system. Being a moving system, the net has no starting-point, no original term. Each topic is potentially the beginning. (Myerson and Rydin 1996, 37)

Moving from this theoretical background and from the notion of environet, this book scrutinises a selection of five texts from the environet, which were drawn from the websites of the five environmental organisations and agencies Forestry England (https://www.forestryengland.uk/), Greenpeace International (https://www.greenpeace.org/international/), National Park Service (https://www.nps.gov/index.htm), Navdanya International (https://navdanyainternational.org/) and World Wide Fund for Nature (WWF) (https://wwf.panda.org/) (see Sect. 1.2). Ecological matters are currently being discussed in an extensive array of texts belonging to different text-types (for instance, fictional and non-fictional prose; official documents; social media; apps) and written by different author-types (such as environment writers; public figures; social media users; community, non-governmental and intergovernmental organisations). This book, however, explores texts belonging to the same text-type, namely websites, and authored by the same author-type, namely environmental organisations or agencies; in this way, the linguistic data and the results emerging from this analysis are more directly comparable.

Furthermore, taking the idea of marker word or environment word as a starting point, this book examines five carefully-selected terms—nature, environment, ecosystem, ecology, sustainability—and their occurrences in the five environmental websites listed above. These five marker words were chosen for investigation considering their connotative abundance, their ideological and emotive many-sidedness and their being loaded with associative meanings; these attributes force the analyst to evaluate precisely what is usually meant by them in ecological texts and discourses. As a result, these terms are keywords in the fields of ecostylistics and ecolinguistics, but originate in and have been defined by the subject areas of ecology and environmental studies (Allaby 2015; Bailey 2004; Collin 2011; Editors of Salem Press 2004; Kemp 2002; Miller and Spoolman 2019; Smith and Smith 2015). Inquiry into the language of nature and the environment necessarily overlaps disciplines and disciplinary theories and knowledge, and requires an explicitly transdisciplinary frame. Consequently, as part of unbiased research, in this book these five marker words are delineated from two complementary perspectives: the linguistic standpoint of lexicology and the naturalistic standpoint of ecology and environmental studies. More precisely, on the one hand, I consulted the Oxford English Dictionary (OED Online 2022), and on the other hand, recent ecology and environment dictionaries and textbooks in their latest editions or latest reprints. All the definitions of these terms are critically appraised here from the viewpoint of ecostylistics, and judged against the ecosophy, or normative ecological philosophy, of this research and against its seven norms (Naess 1995a, 8; Stibbe 2021, 11–16; see Sect. 1.3).

In the ecostylistic research on the five marker words in the five environmental websites presented here, the theoretical framework of ecolinguistics combines with the methodology of stylistics (for discussion and complete reference, see Sect. 2.3). The analytical field of ecostylistics was selected among others as it links the ecological concerns of ecolinguistics to the rigorous reading and interpretation of text and discourse characterising stylistics, particularly the stylistic paradigm adhered to here and defined by Carter (2010, 58–61) as "practical stylistics" or "steam stylistics" (see Sect. 2.2). Consequently, one of the objectives of this book is to confirm that stylistics, which has always been an interdisciplinary area, can be successfully related to ecolinguistics too. Moreover, this book satisfies the need for stylistics to reinforce its social orientation and scholarly engagement in environmentalism and ecological issues, which are now of topical interest.

With regard to ecolinguistics, research in this field so far has mostly scrutinised texts and discourses about nature which, in Stibbe's terminology (2021, 22–26), are either ambivalent, i.e. they only partially encourage people to protect the earth, or destructive, i.e. they encourage people to destroy the earth (see Sects. 2.1.3 and 2.1.4). Ambivalent discourses give implicit precedence to our human economic priorities and higher economic growth, which clashes with the environment-oriented causes and eco-friendly projects the same texts explicitly underwrite and champion. Ambivalent and destructive discourses alike express anthropocentric views: they principally value nature not in its own right, but in terms of the advantages and benefits it may offer us humans.

Detecting beneficial discourses and beneficial discursive strategies encouraging people to protect the earth (Stibbe 2021, 26–30; see Sect. 2.1.2) is hence an undeveloped but essential research area in both ecostylistics and ecolinguistics. These disciplines should more often go beyond the critical assessment of ambivalent and destructive discourses, and inspire the circulation of new beneficial discourses; this is in line with the field of positive discourse analysis, introduced by Martin (1999, 2004, 2006; see also Bartlett 2009 and 2017, and Hughes 2018) and adopted within ecolinguistics by Stibbe (2018). This book is accordingly written in response to the work identifying and exposing ambivalent and destructive discourses only; in fact, one of its objectives is to fill a research gap and to make an analytical contribution to the undeveloped ecostylistic and ecolinguistic research area exploring beneficial discourses.

As a result, in accordance with the theoretical frameworks and methodologies of ecostylistics and ecolinguistics, the main research purpose of this book is twofold:

1. To identify beneficial discourses in the environet, in particular in the texts of the five environmental websites, and to examine the beneficial ecostylistic strategies utilised to produce them. Beneficial discourses and beneficial ecostylistic practices are those assessed against the ecosophy of this research and found to be in line with at least one of its seven norms. These discourses and strategies celebrate the centrality of life and all life forms, convey new ways to conceptualise the planet and to interact with it, and urge their addressees to conserve nature and the environment; in sum, it is the beneficial ecological worldviews relayed by these strategies that we humans must found our societies and our relations with the natural world on.
2. Once these strategies have been identified and studied, to promote a beneficial use of language supporting more successful propagation and popularisation among the general public of beneficial mindsets. This should foster not only academic debate, but also ecological and social debate and, finally, ecological and social change.

Although the texts from the five environmental websites investigated in this book are mostly beneficial, I hypothesise that a small number of phrases and clauses might feature ecostylistic characteristics which can be called ambivalent or, worse, destructive. If this hypothesis is supported, I will point out those ambivalent or destructive characteristics, reword the phrases or clauses in a beneficial way, and discuss the new ecostylistic choices and practices.

In order to achieve its main research purpose, this book is organised as follows. This introductory Chapter 1 presents the five websites authored by environmental organisations or agencies scrutinised in the five analytical Chapters 3–7 (Sect. 1.2) and outlines the beneficial agendas and activities of these organisations and agencies. The introductory Chapter 1 also treats the notion of ecosophy (Sect. 1.3): this is one of the primary terms deployed in this book and is defined from the disciplinary viewpoint of ecolinguistics. The theoretical Chapter 2 covers the field of ecostylistics (Sect. 2.3), along with those of ecolinguistics (Sect. 2.1) and stylistics (Sect. 2.2). In addition, Sect. 2.1 on ecolinguistics deals with the idea of discourse in this area (Sect. 2.1.1) and delineates beneficial discourses (Sect. 2.1.2) versus ambivalent and destructive discourses (Sects. 2.1.3 and 2.1.4). Section 2.2 on stylistics also discusses the major stylistic techniques occurring in the texts explored in this book and coming to light by means of the methodology of practical stylistics: foregrounding (Sect. 2.2.1), point of view (Sect. 2.2.2) and metaphor (Sect. 2.2.3). In accordance with the tradition and approach of practical stylistics, these sections are complemented by the Appendix to this book (Chapter 9), which supplies three checklists to investigate foregrounding, point of view and metaphor.

Following the theoretical Chapter 2, each of the five analytical Chapters 3–7 examines one of the five marker words: nature (Chapter 3), environment (Chapter 4), ecosystem (Chapter 5), ecology (Chapter 6) and sustainability (Chapter 7). The five analytical chapters share the same structure; all of them are divided into two main sections. In compliance with practical stylistics, the first section, "Definition and Evaluation", investigates the meaning of the marker word under scrutiny as defined from the lexicological standpoint of the Oxford English Dictionary (OED Online 2022) and from the naturalistic standpoint of ecology and environmental studies. The first section also evaluates the definitions of the marker word against the ecosophy of this research and its norms, and demonstrates whether these definitions are beneficial, ambivalent or destructive; if they are ambivalent or destructive, they are rephrased in a beneficial way. The definition of marker words from two different perspectives and their evaluation from an ecostylistic viewpoint needs to be careful and detailed, as does a practical stylistic study of text and discourse (see below). Therefore, although further marker words would merit ecostylistic investigation (see, among others, "biodiversity"), the focus of this book was circumscribed to five of them for reasons of space.

The second section of the five analytical chapters, "'[Marker Word]' in [Environmental Organisation or Agency]'s Words", is dedicated to the marker word in the text of one of the five environmental websites. Hence, each of the five analytical chapters considers one of the five marker words as it can be found in one of the five environmental websites. The second section is, in turn, organised into two sub-sections. The first sub-section, "Ecostylistic Analysis", explores a short sequence from an environmental website including the marker word by undertaking a close ecostylistic interpretation based on practical stylistics and on the stylistic mechanisms of foregrounding, point of view and metaphor. Comparatively short sequences are examined, since the type of accurate linguistic and discursive reading typifying practical stylistics can more effectively be applied to relatively short texts.[2] The second sub-section, "Summary and Evaluation", provides an assessment of the value-system of the sequence emerging from the ecostylistic study; this value-system is compared with the ecosophy of this research, which confirms whether the discourse communicated by the sequence is fully beneficial, or rather shows a small number of ambivalent or destructive features requiring beneficial rephrasing. Because all the five analytical chapters define and evaluate a marker word, and treat and assess its occurrence in one of the five environmental websites, the chapters have the same analytical structure and are explicitly linked to one another. Furthermore, in-chapter references to previous analytical chapters and analytical chapters to come are made, both to the topics of the texts under investigation and to the functions of the ecostylistic traits recognised in them.

Given the five marker words and the five environmental websites, the concluding Chapter 8 summarises the contents of this book and the findings presented in the second sections of the five analytical Chapters 3–7. It pulls all the central threads in the five analytical chapters together by making comparisons between them and by referring back to the main research purpose of this work: to scrutinise beneficial discourses and to highlight beneficial ecostylistic devices. To this end, Chapter 8 is divided into four sections. Section 8.1 reminds the reader of this book that underpinning this research are the ecosophy and its seven norms, the theoretical framework and methodology of ecostylistics, the definitions of the five marker words and the results outlined in the second sections of the five analytical Chapters 3–7. As this ecostylistic research is conducted via the stylistic practices of foregrounding, point of view and metaphor, each of the three following Sects. 8.2–8.4 is devoted to one of these stylistic practices. These three sections show how these stylistic techniques are utilised to discuss the ecological questions triggered by the use of the five marker words in the five environmental websites; in other words, how these stylistic mechanisms are deployed to depict our human connections with nature and the environment in a beneficial way. Section 8.5 offers some concluding remarks on these beneficial strategies.

1.2 Data and Methodological Notes

As mentioned in Sect. 1.1, this book explores the instances and uses of the five marker words "nature", "environment", "ecosystem", "ecology" and "sustainability" on the websites of the five environmental organisations and agencies Forestry England (https://www.forestryengland.uk/), Greenpeace International (https://www.greenpeace.org/international/), National Park Service (https://www.nps.gov/index.htm), Navdanya International (https://navdanyainternational.org/) and World Wide Fund for Nature (WWF) (https://wwf.panda.org/). Why were these organisations and agencies selected for investigation here? They have dissimilar missions and do not share the same scope of influence, some being international and others being country-specific. However, all of them, from their various perspectives, belong to the continuum of the ecological and environmental movement, with its infinite number of gradations. Moreover, all their websites are written in English, including those of country-specific Forestry England and National Park Service; as this is the lingua franca of online communication worldwide, all the websites are equally accessible around the world. Thus, the five environmental websites of these organisations and agencies all convey beneficial stylistic devices and discourses, and contribute to producing the collection of texts composing the global social and cultural environet.

The first environmental website, in alphabetical order, to be analysed in this book is that of the national agency Forestry England (https://www.forestryengland.uk/). The agency is responsible for growing and maintaining over 1,500 English forests, more than any other organisation in the country. In Forestry England's ecosophy, forests are managed to the advantage of people, nature and the economy; indeed, the agency has three aims. Firstly, it improves people's quality of life by providing us with recreation and physical and mental wellbeing, such as those given by watching wildlife and exercising or playing among the trees. To name just a few activities encouraging people to become involved with their local forests, it has built more than 1,800 miles of trails for walking, running and cycling, and has formulated several programmes catering for a number of needs and tastes, from sportspeople and children to music and art lovers. Secondly, Forestry England offers wildlife native habitats in diverse and healthy forests, thereby preserving thousands of animal and vegetable organisms; in addition, it is restoring lost species to their former spaces and conditions, and reclaiming ecosystems essential to life. It is one of the largest English managers of forests, woodlands, moors, heathlands and bogs recognised as Sites of Special Scientific Interest, where extremely rare wildlife can not only survive but thrive. Thirdly, the agency supplies the economy with the country's largest amount of sustainably-harvested timber, thus supporting the British forestry industry and local economies around the country. In short, in its ecosophy Forestry England regards forests as sustaining life on earth, given that, among other natural processes, they store carbon and contain floods; with regard to us humans, forests furnish us with fresh air and healthy spaces, namely wellbeing and leisure. For these reasons, the agency makes forests flourish through careful management and planning, and plants a wide variety of resilient tree species: they make various kinds of food and nectars available to wildlife and will be enjoyed in the future too (https://www.forestryengland.uk/about-us; https://www.forestryengland.uk/managing-people; https://www.forestryengland.uk/managing-nature). In this book, a sequence from Forestry England's website (Text 3) is employed to examine the use of the marker word "ecosystem" (Sect. 5.2).

The website of the international organisation Greenpeace is the second environmental website studied here (https://www.greenpeace.org/international/). Founded in 1971, the organisation consists of twenty-six independent national or regional organisations in more than fifty-five countries, in addition to the coordinating body Greenpeace International; it is this body's website that is investigated in this book. In Greenpeace's ecosophy, our planet is regarded as our home: nature is not considered to be a source of income, but a source of life and beauty to be valued per se. Accordingly, established narratives of the way of the world accepting and preserving the status quo are rebuffed and replaced with novel discourses of a greener, fairer and more peaceful world. To be more precise, the organisation's mission is "to ensure the ability of the earth to nurture life in all its diversity" (https://www.greenpeace.org/international/explore/about/values/). This implies safeguarding all forms of biodiversity, stopping abuse and pollution of the planet's land, air, oceans and freshwater, putting an end to all nuclear risk, supporting peace, nonviolence and disarmament worldwide. In line with the organisation's ecosophy, different global environmental problems can be exposed by promoting concrete solutions and taking cooperative action against ecological destruction and the social and political systems threatening the environment. Greenpeace's activists are trained in nonviolence and are personally accountable for what they do. Their various activities range from planting trees and voting for green candidates to participating in protest marches against governments and corporations and campaigning on the organisation's fleet. In order to ensure its independence and to avoid external influence, Greenpeace only accepts donations from individuals and grants from private foundations to fund its campaigning work. The organisation's confrontational campaigns are not only peaceful, but also stylistically creative and communicatively powerful, and clearly relay its ecosophy and ethical principles (https://www.greenpeace.org/international/explore/about/values/). In this book, the use of the marker word "environment" (Sect. 4.2) is scrutinised by means of a sequence from Greenpeace International's website (Text 2).

The third environmental website explored here is that of the US federal agency National Park Service (https://www.nps.gov/index.htm). The agency is a bureau of the United States Department of the Interior with headquarters in Washington, DC; since its creation in 1916, it has been mandated to preserve, protect and manage the country's national parks, which now employ approximately 20,000 staff and are visited by more than 318 million people a year. The National Park System "includes 423 areas covering more than 85 million acres in every state, the District of Columbia, American Samoa, Guam, Puerto Rico, and the Virgin Islands. These areas include national parks, monuments, battlefields, military parks, historical parks, historic sites, lakeshores, seashores, recreation areas, scenic rivers and trails, and the White House"; they also embrace some of the rarest and most endangered ecosystems and habitats in the world and unique combinations of wildlife and plants, whose integrity is maintained by the agency (https://www.nps.gov/aboutus/faqs.htm). Underlying the National Park Service's ecosophy is the concept of national park. This establishes the belief that the nation should take care of and be answerable for public lands, and goes beyond protecting one-of-a-kind natural scenery: it considers wildernesses to be the heritage of the entire nation, which benefits less from private businesses utilising the physical environment for their own ends than from a rewarding personal experience in nature (https://www.nps.gov/yell/learn/historyculture/modernmanagement.htm). The National Park Service works with Native Americans, local governments, nonprofit-making organisations, businesses and individual citizens to put new life into communities, conserve the history of towns and counties, pay tribute to their heritage and provide recreation opportunities for adults and children alike in nearby parks throughout the country. The agency's mission is to safeguard and maintain unspoilt the natural and cultural heritage of the National Park System, i.e. its wildlife, vegetation and scenic, recreational, historical and archaeological values, so that this generation and those to come can benefit and get pleasure, inspiration and education from it (https://www.nps.gov/aboutus/index.htmhttps://www.nps.gov/aboutus/faqs.htm). In this book, a sequence from the National Park Service's website (Text 4) is utilised to analyse the use of the marker word "ecology" (Sect. 6.2).

The website of the international movement Navdanya International is the fourth environmental website examined here (https://navdanyainternational.org/). Navdanya originated in India, around thirty years ago, in the work and ecosophy of the physicist, philosopher and environmental activist Vandana Shiva, with the goal to defend small farmers globally and their right to seed and food sovereignty. Navdanya initiated and led the movement of seed saving and seed freedom, which started out to react against the gradual destruction of agricultural biodiversity and the newly-introduced genetically modified organisms (GMOs) and patents on seeds. It strives and campaigns against biopiracy and the patenting of traditional ecological knowledge by environmentally-unconcerned multinational corporations, and has won cases connected with neem, wheat and basmati rice in India. Following its ecosophy, the movement advocates a culture of food for health, that is to say, a major conceptual change in agricultural and economic practices. The hegemonic selfishness, materialism and consumerism currently characterising our societies must be replaced by principles like the common good, ecological accountability, economic equality, solidarity, a culture of peace and a restored sense of community. Navdanya thinks of biodiversity as a gift to be shared by all and fights to accomplish its objectives by safeguarding, replenishing and rejuvenating it as it was handed down by our ancestors (https://navdanyainternational.org/about-us-navdanya-international/). One of the principal activities it promotes is researching and disseminating agroecology, or the "holistic study of agro-ecosystems, including all environmental and human elements[,] that focuses on the form, dynamics and functions of their inter- and intra-relationships" (https://navdanyainternational.org/key-issues/agroecology/). The movement's research has proven that agroecology can enhance our human nutrition and health and raise small farmers' incomes; at the same time, it can reclaim soil, water and biodiversity and improve adaptation to a changing climate. In 2011, Navdanya International was established in Italy to further the movement's ecosophy and activities, and to preserve nature, biodiversity, farmers' rights and traditional ecological knowledge and culture worldwide (https://navdanyainternational.org/about-us-navdanya-international/). In this book, the use of the marker word "sustainability" (Sect. 7.2) is studied through a sequence from Navdanya International's website (Text 5).

The fifth and last environmental website investigated here is that of the international conservation organisation World Wide Fund for Nature (WWF) (formerly, and in Canada and the United States, World Wildlife Fund) (https://wwf.panda.org/). The organisation was founded in 1961 and is now active in almost 100 countries. In WWF's ecosophy, nature is the system providing the essentials of all life, including our human life, so much so that we cannot manage without it: if nature is lost, none of us will be able to survive. Hence, the organisation works to support and reclaim the environment and natural habitats in the interests of all human animals (henceforth humans), nonhuman animals (henceforth animals) and plants. WWF's mission is to cease the deepening crisis of nature loss and dilapidation. Its objectives and plan of action around the globe centre on: (1) safeguarding the natural world and fighting the all-pervasive tendency of biodiversity loss and mass extinction of wildlife now affecting the planet; (2) making certain that renewable natural materials are employed in a sustainable way in our production and consumption practices, particularly in our food production, which is one of the most urgent threats to nature; (3) advocating cutting consumption and waste and reducing pollution by confronting the most widespread triggers for nature loss, viz. ecologically and socially unjust financial systems, governance systems and markets. WWF's activities are determined and given a direction by its tireless dedication to cooperation with indigenous peoples and local communities and with other environmental organisations, which has brought about several close associations worldwide (https://wwf.panda.org/discover/about_wwf/). WWF's ecosophy is reflected in *Living Planet Report*, its principal publication: appearing every two years, it is a thorough examination of the tendencies in the condition of the earth and in the biodiversity of animals, plants and their habitats worldwide (see WWF 2020a, the thirteenth and latest edition of the report). In this book, a sequence from WWF's website (Text 1) is deployed to scrutinise the use of the marker word "nature" (Sect. 3.2).

The activist ecosophies, work and occupations of these five environmental organisations and agencies are introduced and outlined in their websites. The texts of the five environmental websites are studied here by investigating the occurrences of the five marker words in beneficial short sequences drawn from the websites. The occurrences and sequences were collected in autumn 2021, after the 2019 COVID-19 pandemic outbreak and before the 2021 G20 Summit in Rome (Italy) and the 2021 United Nations Climate Change Conference of the Parties (COP26) in Glasgow (United Kingdom); they were decided on as follows. Firstly, I accessed the search tool on the homepages of the five environmental websites and entered one of the five marker words at a time. Forestry England's and National Park Service's search tools sort the results automatically, apparently by relevance (from the most relevant to the least relevant). Navdanya's search tool automatically sorts the results by date (from the newest to the oldest). Greenpeace's and WWF's allow the website user to sort them by either relevance or date; accordingly, I arranged them by relevance. Secondly, I opened the webpages resulting from the search and looked for instances of the marker word under scrutiny through the browser Find tool. Thirdly, I selected sequences where the marker word occurred mainly as a noun (e.g. "nature"), not as an adjective or adverb ("natural" or "naturally"); moreover, the noun must not be premodified by classifying adjectives restricting its meaning ("urban nature"), neither must the noun be a premodifier itself ("nature loss") or a constituent in a compound word ("nature-based"). The final choice was qualitative: I decided on sequences which, during a first, but careful reading, seemed to signal a beneficial ecosophy mostly aligning with that of this research and to display a number of beneficial ecostylistic aspects. Whenever possible, I preferred sequences comprising more than one instance of the marker word, or sequences explicitly or implicitly furnishing a definition or description of it. By following this procedure, I selected five sequences whose word counts range from 174 words (from National Park Service's website, on the marker word "ecology") to 273 words (from Greenpeace's website, on the marker word "environment"). As mentioned above and in Sect. 1.1, each of the five sequences derives from one of the five environmental websites; each sequence is devoted to one of the five marker words and is explored in one of the five analytical Chapters 3–7. The ecosophy of this research, which influenced the choice of the five sequences, is presented in Sect. 1.3 below.

1.3 The Ecosophy of This Research

Underlying the concept of ecosophy are the notions of "stories" and "stories-we-live-by", which Stibbe (2021, 6) defines as follows: "*Stories* are cognitive structures in the minds of individuals which influence how they think, talk and act. *Stories-we-live-by* are stories in the minds of multiple individuals across a culture". Instances of stories are anthropocentrism, individualism, economism, progress and competition (for the last two stories, explored as "feedback metaphors", see Larson 2011, Chapters 2 and 3). Stories are not usual or traditional narratives with a plot unfolding in the form of a novel or film, or facts related by an addresser to an addressee; they are not directly stated or immediately recognisable in text and discourse. They instead underpin, pervade and discipline our current unsustainable culture by structuring the hegemonic texts and discourses in industrial societies and the mainstream ways of speaking, writing and expressing ourselves in our everyday life, from institutional contexts to interactions with friends. We take stories for granted as a necessary background and experience them as actual facts; as a result, they provide specific frameworks for reading and understating our human existence.

Stories are therefore cognitive patterns embedded in the brain and way of thinking of manifold people across a given culture. Since they are pervasively told and shared in the way of life of a society, they are reckoned to have been ethically and conceptually legitimised. This entails that they have a deep social and cultural impact on our human actions and human-caused problems, like the environmental crisis, biodiversity loss and injustice. As such, they are dangerous and must be uncovered, critically addressed and challenged. If they are involved in and lead to social inequality and ecological devastation, they must be resisted; if they urge us to feel concern and consideration for the ecosystems sustaining life, they must be promoted. Nevertheless, this is not easy: we are subjected to their risks without deliberately choosing these stories and without realising that they are mere constructs (Stibbe 2014, 117; Stibbe 2021, 2–6).

Examples of dominant stories-we-live-by are the following:

1. The story of human centrality: nature and animals merely depicted as a free supply of available assets detached from us humans, which we humans can take advantage of (Kingsnorth and Hine 2009);
2. The story of human exceptionalism: we humans as unique superior non-animal beings characterised by the dissimilarities, not the shared features, between us, other animals and the natural world (Plumwood 2007);
3. The business-as-usual story: unrestrained economic growth, which has been thought of as the basic economic model and the primary objective of society and politics for a few hundred years (Macy and Johnstone 2012, 15; see also Robertson 2014, 54);

4. The amassment of unneeded objects as the means to self-improvement (Stibbe 2021, 3);
5. Prosperity and advancement imagined as simple material benefits and as the consistent introduction and development of new technologies (Stibbe 2021, 3).

According to Korten (2006, 248), industrial civilisation and Western imperial societies are founded on four essential stories-we-live-by with profound ecological consequences:

1. The "prosperity story" encourages veneration for wealth and possessions;
2. The "biblical story" is not centred on this world, but rather on the afterworld;
3. The "security story" maintains relations of oppression by establishing the army and civil forces;
4. The "secular meaning story" reductively regards life as mere physical substance and mechanical forces.

As stated by this author, not only do these four stories preserve inequality, they also bring about ecological devastation, lasting estrangement from nature and an environmentally unsustainable way of life.

The aim of unmasking the stories-we-live-by is to make them accessible to dispute and argument; more precisely, it is to explore whether they are fit for and function in the present ecological state of affairs, or whether novel and more appropriate stories are required. This is determined by the analyst's ecosophy and evaluative framework, namely by whether or not the sort of society the analyst wishes for is being shaped by the stories under examination. Ecolinguistic investigation is normally founded on certain ecological principles, stances and courses of action; it searches for stories advocating them, which are assessed positively and championed by the analyst in their studies for practical use by the general public. These principles, stances and courses of action are also common to the most academically activist scrutinies in ecostylistics (see Sect. 2.3). They generally include the following priorities: cooperation versus competition, sharing versus greed, peace versus violence, ecological sustainability versus ecological destruction, we humans connected with nature and animals versus we humans separate from nature and animals (Stibbe 2021, 12); in short, what was called, comprehensively, "socioecological sustainability" (Larson 2011, 17).

As instances of ecological discourse, ecostylistic and ecolinguistic analyses are characterised by evaluative language founded on an articulate interaction between moral and aesthetic criteria, assessing and critically commenting on ways of life, thinking and acting, and outlining our human responsibilities and duties to the planetary ecology. Harré et al. (1999, 179–182) recognised a variety of moral discourses in environmental writing and speaking, or Greenspeak, and elaborated a "taxonomy of moralities", or three main moral arguments: rights-based arguments, arguments based on virtues and vices, and utilitarian and intrinsic values. In line with these arguments, ecology-oriented discourse increases the extent of the moral sphere and the range of morally-protected entities. The reasons for this moral extension do not originate from human needs or utilitarian arguments, but from the richness and diversity of species and all life forms being an end in themselves; accordingly, what is also extended is advocacy rights for all living beings.

The language of evaluation and moral and aesthetic criteria also typify ecosophies. The term "ecosophy" (a blend of "ecological philosophy") was coined by the philosopher Arne Naess (1995a, 8) and defined as follows:

> By an *ecosophy* I mean a philosophy of ecological harmony or equilibrium. A philosophy as a kind of *sofia* [or] wisdom, is openly normative, it contains *both* norms, rules, postulates, value priority announcements *and* hypotheses concerning the state of affairs of our universe. Wisdom is policy wisdom, prescription, not only scientific description and prediction. (Naess 1995a, 8)

That is to say, an ecosophy is a statement of ecological values and assumptions, a system of philosophical tenets based on and highlighting environmental concerns (see also Naess 1995b). An ecosophy is an ethics of ecological concord; as such, it is explicitly normative and prescriptive by establishing rules and standards and by implying value priorities. An ecosophical system is conveyed verbally as a set of sentences with various descriptive and prescriptive functions. Taking into account the large extent of the pertinent ecological contents and normative—social, political, ethical—contents of the ecosophy, a description of such a complex and flexible system can inevitably be only averagely definite and exact: actually, "It is the global character, not preciseness in detail, which distinguishes an ecosophy" (Naess 1995a, 8).

An ecosophy is a moving, ever-changing structure: the items in an analyst's ecosophy are subject to change because of the variations in both the analyst's key interests and considerations and the ecological topics they may decide to emphasise in their distinct works. Consequently, explorations of text and discourse cannot be founded on a single ecosophy which is assumed to be accurate on all occasions. Each analyst should investigate the broad array of available ecosophies developed by researchers, select the most suitable to their ecological and social knowledge, or elaborate their own. The chosen ecosophy must be feasible from a scientific viewpoint, seem practically reasonable and acceptable, and adjust to the existing scientific facts and data (Stibbe 2021, 14).

According to Stibbe (2021, 12–13), ecosophies can be fashioned by relying on several ecological worldviews, all of which are inclined to be based on three ranges of variables. The first range runs from anthropocentrism to ecocentrism, i.e. from a human-centred value-system to a life-centred value-system. The second set of variables ranges from neoliberalism to socialism, localism and anarchy. The third range of variables runs from optimism to pessimism. The three ranges broadly support each other: for instance, ecocentrism and pessimism are apt to typify political theories advocating thorough and far-reaching change, whereas anthropocentrism and optimism are usual characteristics of neoliberal paradigms. Similarly, Myerson and Rydin (1996, 3–4, 26–27) comment on the ecological debate between the campaigning intellectual Norman Myers and the academic economist Julian Simon: the former alarmingly and insistently sees danger and crisis in a bounded world, while the latter reassuringly and energetically sees growth and industrial production in an expandable world. Stibbe (2021, 13–14) briefly describes a number of ecological philosophical ideologies aligning along the three ranges above, from the most politically conservative cornucopianism (Ridley 2010) to the most radically ecocentric and pessimistic Voluntary Human Extinction Movement (VHEMT 2014).

As conceived of and applied in this book, ecostylistics is founded on ecolinguistics and stylistics (see Sect. 2.3), mainly on the way the former discipline is being developed by Arran Stibbe and by many of the members of the International Ecolinguistics Association; accordingly, the ecosophy of this ecostylistic research is also founded on that presented in Stibbe (2021, 14–15: see also Stibbe 2018). The ecosophy of this research consists of the following norms:

Ecosophy in One Word: Living!

1. **Valuing Living:** The exclamation mark in "Living!" is used to signal that this is a normative principle and a moral imperative; in other words, living must be considered of the highest worth and importance, paid tribute to, revered and asserted strongly. The ecosophy of this research is ecocentric and founded on intrinsic tenets and attitudes. It takes into account the ecosphere as a whole, namely all human, animal and vegetable species living there, and praises their interpenetration and mutual connection; if this interconnected state is balanced and all species live in harmony, living becomes living well, which is the object of norm no. 2: Wellbeing. The ecosophy also advocates care, respect and deep concern for all animal and vegetable organisms, regardless of the helpfulness of nonhuman species for the interests and objectives of our human species. That is to say, all humans, animals and plants are comprised in the ecosophy, given that it has been remarked that living is an overriding priority for all of us, and that we are always ready to do anything in our power to keep on living, whether in a deliberate way, by instinct or automatically. This is the reason why, in this book, the noun "resources" is never used to refer to the nonhuman world or more-than-human world (Abram 1996) and to its biotic and abiotic components, for instance animals, plants, air, water and soil; such terms as "materials" and "processes",

or simply "nature" and "the natural world", are preferred instead (for an ecostylistic analysis of the noun "resources", see Sect. 6.2).

2. **Wellbeing:** The capacity of all human, animal and vegetable species to value living decreases in negative situations where individuals are alive but suffering; "existing" or "surviving" is diverse from "living". Living in good health, in a contented and prosperous condition, enjoying physical and moral welfare, is hence the thrust of the ecosophy. An ideological change must be made to achieve this purpose: we must prefer quality to quantity and wellbeing to economic growth. We must appreciate life and satisfactions requiring simple material or spiritual means, rather than consumerism and ever higher levels of material wealth in our Western industrial societies—what Odum and Odum (2001, 3, 4) labelled our "unessential baggage" and "useless diversions". In practice, essential conditions for living well are that we humans satisfy our vital and spiritual needs and that wellbeing and flourishing be priorities for us; this is because we are most unlikely to take any serious steps or drastic course of action to tackle ecological challenges which would affect adversely our personal profit, benefit or advantage. Wellbeing, however, must have relevance for all human and more-than-human organisms; therefore, this book, which identifies and promotes beneficial texts and discourses, gives special prominence to those supporting the wellbeing of human, animal and vegetable species alike, and to those fighting extinction and championing their abundance, diversity and conservation.
3. **Now and the Future:** Temporal concepts and dimensions are closely blended with the basic structure of ecological discourse. The past is connected with the future; the present is depicted, on the one hand, as the sum of the outcomes of the past and, on the other hand, as the required springboard to the events of the future (Harré et al. 1999, 7; see also 8–12 and Chapter 6). For this reason, the time range of the ecosophy is unlimited: it embraces the ability to live well in the present as well as in the future; that is, it incorporates both present and future generations living with a high level of wellbeing. The mainstream definitions of the noun "sustainability" (see Sect. 7.1) mainly focus on the state of affairs in the decades to come; nevertheless, ecological problems are with us now, and drastic action

and measures to save the planet must be taken immediately, not in the future. Accordingly, in this book, the term "environmental crisis", taken from the title of an article by the sociologist Saroj Chawla (1991), has been chosen over similar terms like "climate change" or "global warming". Its meaning is broader and relays more urgency than the other options: in this context, the noun "environment" (hence the adjective "environmental") is a hypernym of the nouns "climate" and "warming"; the value-laden noun "crisis" denotes a time of trouble and danger, contrary to the value-neutral noun "change" (for the issue of the terminology for the phenomenon, see Penz 2018).

4. **Care:** The fact that wellbeing must be high only for us humans entails that we kill impressively large numbers of animal and vegetable organisms or that we harm their lives so that we can carry on living well ourselves. The ecosophy bears this moral issue in mind by not regarding the animals and plants we kill and damage as subordinate entities, mere resources or as having no intrinsic value, and by introducing the concept of care, through which we build a subjective, not objectified, relation to them. Care encompasses the notions of empathy, regret and gratitude. In this ecostylistic and ecolinguistic context, empathy is the consciousness of our negative influence on other organisms; regret means reducing that influence and damage; gratitude is our moral obligation to give something in return to the species and the ecological structures sustaining us. However, a key argument in the ecosophy is high regard for all animals and plants, because they are ends in themselves. As a result, as mentioned in norm no. 2: Wellbeing, this book advocates primarily the beneficial ecostylistic strategies championing the welfare of all organisms, whether human, animal or vegetable. In order to emphasise our human responsibility and accountability for the ways we use the more-than-human world, in this book the collective nouns "humanity" or "humankind" are not employed, given that they denote, in the singular, more than one individual and, as such, appear to designate somewhat abstract referents. I have preferred the noun and adjective "human", usually preceded by the first-person plural pronouns "we", "us" and "our"; I utilise these pronouns inclusively, viz. comprising

the addresser (myself as the author of this research), the addressee (the book readers) and all the other humans.

5. **Environmental Limits:** The ecosophy is underpinned by a copious quantity of contemporary and up-to-date evidence supported and circulated by the vast majority of the scientific community, such as that presented in the Intergovernmental Panel on Climate Change (IPCC)'s assessment reports and special reports (see, among others, IPCC 2019a, 2019b, 2021, 2022a, 2022b), World Wide Fund for Nature (WWF)'s *Living Planet Report 2020* (WWF 2020a) and the *Living Planet Index* (WWF 2020b) the report is based on. Such evidence argues that the capacity of ecological structures to sustain life, consequently wellbeing, into the years and decades to come is affected adversely by us humans overusing goods, materials and energy and generating waste: these behaviours will soon hamper humans, animals and plants in their efforts to live well and the natural world to restore itself. Accordingly, the extraordinary rates of consumption and waste production worldwide, particularly in the most industrialised countries, put the ecosphere in grave danger and affect it severely: they must be cut down urgently and extensively and maintained within environmental limits. Moreover, in accordance with the ecosophy of the deep ecology movement (Naess and Sessions 1995; see Sect. 2.1.2), a drastic stabilisation and reduction of our human population is necessary for human life and, above all, for more-than-human life to flourish. In fact, currently, we humans are interfering with the lives of nonhuman organisms inordinately, and their conditions are getting more and more difficult (see also Odum and Odum 2001 and Odum 2007).
6. **Social Justice:** The protection of the environment is interconnected with the actualising of social justice. At the present time, millions lack the means to live, never mind to live well. These circumstances will worsen shortly, for consumption standards will decline throughout the world, either intentionally or when we humans have completely consumed nature and natural materials. In addition, poor, disempowered and involuntarily displaced peoples mainly from the Global South, along with the ecosystems they are a part of, are already vulnerable; their vulnerability is aggravated by the phenomenon of

slow violence (Nixon 2011), or the threats occurring progressively, whose damaging consequences will be fully and dramatically visible in the long run, and which are thus disregarded in the short run, like toxic drift, deforestation and oil spills. Hence, if every human, animal and vegetable species is to live well, we humans from the Global North will be required to create a profoundly different state of affairs from the present one. We must change our economic, technological and ideological policies; replace individualism and economic competition with global cooperation and policies based on mutual understanding; respect and strengthen local cultures, characteristics and customs; eliminate affluence, luxury and waste; reduce inequality; and distribute global wealth and natural materials more fairly (see Odum and Odum 2001).

7. **Adaptation:** The path currently followed by the countries affected by industrialism and overdevelopment has already led nature to notable ecological devastation and will inevitably lead to further ruin. The highest point of material prosperity of our global social and economic development is right in front of us, and so is the resulting downturn and eventual breakdown of it. Since the present ecological, social and economic systems are failing, we humans must recognise that our lives will change and respond to this state of affairs by thinking ahead and making the necessary plans for our future. We must therefore adjust to new ecological conditions, raise our resilience to future alteration and evolve new social and economic systems by reorganising the present ones on more sustainable ideals and objectives (for the new meaning of "moving forward" taken on by the term "resilience", and for discourses of social-ecological resilience, see Plastina 2020). These adjustments must be made to keep on living well and to achieve a new balance with our ecological life-support systems while the planet and the climate are turning unfriendlier and more unsuitable for life (for a detailed model of transition and turndown, see the one developed by the ecologist and general systems theorist Howard Thomas Odum [Odum and Odum 2001]).

(Adapted from Stibbe 2021, 14–15. See also Naess and Sessions 1995; Odum and Odum 2001; Odum 2007).

As mentioned in the introductory Sect. 1.1, this ecosophy provides the touchstone against which to compare and judge the different beneficial discourses relayed by the texts scrutinised in the five analytical Chapters 3–7. Before moving on to these chapters and to exploring the ecological texts and discourses selected for ecostylistic examination, ecostylistics, as well as ecolinguistics and stylistics, is covered in the theoretical Chapter 2 below.

Notes

1. As shown in Sects. 4.1 and 6.1, the meaning of the noun "ecology" is more comprehensive than that of the noun "environment"; nevertheless, in this book, the adjective "environmental" is sometimes employed interchangeably with the adjective "ecological" for reasons of elegant variation.
2. Further research and a future development of the work undertaken in this book could move, as Carter (2010, 65) put it, "from steam to broadband", that is to say, employ corpus-based methods of stylistic scrutiny. For instance, taking this ecostylistic analysis of the five marker words in the five beneficial environmental websites as a starting point, one could explore their occurrences, their uses and the value-systems they convey in one text-type or in different text-types, by examining a large number of texts typical of that text-type or those text-types with a corpus stylistic approach. However, this goes beyond the aims and scope of this research: what I am interested in investigating and evaluating in detail here is the specific value-system of the five beneficial sequences, their ecostylistic strategies and their meaning-making practices.

References

Abram, D. 1996. *The Spell of the Sensuous: Perception and Language in a More-than-Human World*. New York: Vintage Books.

Allaby, M. 2015 [1994]. *A Dictionary of Ecology*, 5th ed. Oxford: Oxford University Press.

Bailey, J., ed. 2004. *The Facts on File Dictionary of Ecology and the Environment*. New York: Market House Books.

Bartlett, T. 2009. Towards intervention in positive discourse analysis. *Applied Linguistics Methods: A Reader*, eds. C. Coffin, T. Lillis, and K. O'Halloran, 133–147, London and New York: Routledge.

Bartlett, T. 2017. Positive discourse analysis. In *The Routledge Handbook of Critical Discourse Studies*, ed. J. Flowerdew and J.E. Richardson, 133–148. London and New York: Routledge.

Carter, R. 2010. Methodologies for stylistic analysis. In *Language and Style: In Honour of Mick Short*, eds. D. McIntyre and B. Busse, 55–68. Basingstoke: Palgrave Macmillan.

Chawla, S. 1991. Linguistic and philosophical roots of our environmental crisis. *Environmental Ethics* 13 (3): 253–273.

Collin, P.H. 2011 [1985]. *Dictionary of Environment & Ecology*, 5th edition. London: Bloomsbury [electronic edition].

Editors of Salem Press, ed. 2004. *Ecology Basics. Pasadena, CA, and Hackensack*. NJ: Salem Press.

Harré, R., J. Brockmeier, and P. Mühlhäuser. 1999. *Greenspeak: A Study of Environmental Discourse*. London: Sage.

Hughes, J.M.F. 2018. Progressing Positive Discourse Analysis and/in Critical Discourse Studies: Reconstructing resistance through progressive discourse analysis. *Review of Communication* 18 (3): 193–211.

Intergovernmental Panel on Climate Change (IPCC). 2019a. *Climate Change and Land: Special Report*. Geneva, Switzerland: IPCC. https://www.ipcc.ch/srccl/. Last accessed April 2022.

Intergovernmental Panel on Climate Change (IPCC). 2019b. *The Ocean and Cryosphere in a Changing Climate: Special Report*. Geneva, Switzerland: IPCC. https://www.ipcc.ch/srocc/. Last accessed April 2022.

Intergovernmental Panel on Climate Change (IPCC). 2021. *Climate Change 2021: The Physical Science Basis: Contribution of Working Group I to the Sixth Assessment Report of the Intergovernmental Panel on Climate Change*, eds. V. Masson-Delmotte, P. Zhai, A. Pirani, et al. Geneva, Switzerland: IPCC. https://www.ipcc.ch/report/ar6/wg1/. Last accessed April 2022.

Intergovernmental Panel on Climate Change (IPCC). 2022a. *Climate Change 2022: Impacts, Adaptation and Vulnerability: Working Group II Contribution to the Sixth Assessment Report of the Intergovernmental Panel on Climate Change*, coordinating lead authors R. Ara Begum and R.

Lempert. Geneva, Switzerland: IPCC. https://www.ipcc.ch/report/ar6/wg2/. Last accessed April 2022.
Intergovernmental Panel on Climate Change (IPCC). 2022b. *Climate Change 2022: Mitigation of Climate Change: Working Group III Contribution to the Sixth Assessment Report of the Intergovernmental Panel on Climate Change*, coordinating lead authors M. Grubb and C. Okereke. Geneva, Switzerland: IPCC. https://www.ipcc.ch/report/ar6/wg3/. Last accessed April 2022.
Kemp, D.D. 2002 [1998]. *The Environment Dictionary*, 2nd edition. London and New York: Routledge [electronic edition].
Kingsnorth, P., and D. Hine. 2009. *The Dark Mountain Project Manifesto*. https://dark-mountain.net/about/manifesto/. Last accessed April 2022.
Korten, D. 2006. *The Great Turning: From Empire to Earth Community*. San Francisco, CA: Berrett-Koehler.
Larson, B. 2011. *Metaphors for Environmental Sustainability: Redefining our Relationship with Nature*. New Haven, CT: Yale University Press.
Macy, J., and C. Johnstone. 2012. *Active Hope: How to Face the Mess We're in without Going Crazy*. Novato, CA: New World Library.
Martin, J.R. 1999. Grace: The logogenesis of freedom. *Discourse Studies* 1 (1): 29–56.
Martin, J.R. 2004. Positive discourse analysis: Solidarity and change. *Revista Canaria De Estudios Ingleses* 49: 179–200.
Martin, J.R. 2006. Vernacular deconstruction: Undermining spin. *DELTA: Documentação de Estudos em Linguística Teórica e Aplicada* 22(1): 177–203.
Miller, G.T., and S.E. Spoolman. 2019 [2007]. *Essentials of Ecology*, 8th ed. Belmont, CA: Cengage.
Myerson, G., and Y. Rydin. 1996. *The Language of Environment: A New Rhetoric*. London and New York: Routledge.
Naess, A. 1995a. The shallow and the deep, long-range ecology movement: A summary. In *The Deep Ecology Movement: An Introductory Anthology*, ed. A. Drengson and Y. Inoue, 3–9. Berkeley, CA: North Atlantic Books.
Naess, A. 1995b. The systematization of the logically ultimate norms and hypotheses of Ecosophy T. In *The Deep Ecology Movement: An Introductory Anthology*, ed. A. Drengson and Y. Inoue, 31–48. Berkeley, CA: North Atlantic Books.
Naess, A., and G. Sessions. 1995. Platform principles of the deep ecology movement. In *The Deep Ecology Movement: An Introductory Anthology*, ed. A. Drengson and Y. Inoue, 49–53. Berkeley, CA: North Atlantic Books.
Nixon, R. 2011. *Slow Violence and the Environmentalism of the Poor*. Cambridge, MA: Harvard University Press.

Odum, H.T. 2007. *Environment, Power and Society for the Twenty-First Century: The Hierarchy of Energy*. New York: Columbia University Press.

Odum, H.T., and E.C. Odum. 2001. *A Prosperous Way Down: Principles and Policies*. Boulder, CO: University Press of Colorado.

Oxford English Dictionary (OED) Online. 2022. Oxford: Oxford University Press. https://www.oed.com/. Last accessed April 2022.

Penz, H. 2018. "Global warming" or "climate change"? In *The Routledge Handbook of Ecolinguistics*, ed. A.F. Fill and H. Penz, 277–292. London and New York: Routledge.

Plastina, A.F. 2020. *Social-Ecological Resilience to Climate Change: Discourses, Frames and Ideologies*. Newcastle upon Tyne: Cambridge Scholars Publishing.

Plumwood, V. 2007. Human exceptionalism and the limitations of animals: A review of Raymond Gaita's *The Philosopher's Dog*. *Australian Humanities Review* 42. http://australianhumanitiesreview.org/2007/08/01/human-exceptionalism-and-the-limitations-of-animals-a-review-of-raimond-gaitas-the-philosophers-dog/. Last accessed April 2022.

Ridley, M. 2010. *The Rational Optimist: How Prosperity Evolves*. New York: Harper.

Robertson, M. 2014. *Sustainability Principles and Practice*. London and New York: Routledge.

Smith, T.M., and R.L. Smith. 2015. *Elements of Ecology*, 9th ed. Harlow: Pearson.

Stibbe, A. 2014. An ecolinguistic approach to critical discourse studies. *Critical Discourse Studies* 11 (1): 117–128.

Stibbe, A. 2018. Positive discourse analysis: Rethinking human ecological relationships. In *The Routledge Handbook of Ecolinguistics*, ed. A.F. Fill and H. Penz, 165–178. London and New York: Routledge.

Stibbe, A. 2021. *[2015]. Ecolinguistics: Language, Ecology and the Stories We Live By*, 2nd ed. London and New York: Routledge.

Voluntary Human Extinction Movement (VHEMT). 2014. *The Voluntary Human Extinction Movement*. http://www.vhemt.org/. Last accessed April 2022.

World Wide Fund For Nature (WWF). 2020a. *Living Planet Report 2020a: Bending the Curve of Biodiversity Loss*, eds. R.E.A. Almond, M. Grooten, and T. Petersen. Gland, Switzerland: WWF. https://f.hubspotusercontent20.net/hubfs/4783129/LPR/PDFs/ENGLISH-FULL.pdf. Last accessed April 2022.

World Wide Fund For Nature (WWF). 2020b. *Living Planet Report 2020b: Bending the Curve of Biodiversity Loss: A Deep Dive into the Living Planet Index*, eds. V. Marconi, L. McRae, S. Deinet, S. Ledger, and F. Freeman. Gland, Switzerland: WWF. https://f.hubspotusercontent20.net/hubfs/4783129/LPR/PDFs/ENGLISH%20-%20TECH%20SUPPLIMENT.pdf. Last accessed April 2022.

2

Ecostylistics, Ecolinguistics and Stylistics: A Theoretical Overview

2.1 Ecolinguistics

This chapter furnishes an overview of ecostylistics, whose theoretical framework and methodology are adopted in this book. The presentation of ecostylistics in Sect. 2.3 is preceded by an outline of the two disciplines inspiring the research in this volume: ecolinguistics (this section) and stylistics (Sect. 2.2). Ecolinguistics is the branch of linguistics connecting the study of language with the environment and the natural world, in their biological or ecological sense. Ecolinguistics investigates the role of language in the interactions among us humans, animals, plants and the environment; it considers language and society as closely joined together with the broader ecological structures embracing and sustaining them and life on earth as a whole. In the most authoritative handbook about the discipline so far and the first exhaustive survey of it (Fill and Penz 2018a), the introductory chapter (Fill 2018, 2) briefly discusses the history of the noun "ecolinguistics". It was seemingly first utilised by the French sociolinguist Jean-Baptiste Marcellesi (1975) in an article about regional languages; subsequently, in a book about linguistic alienation by the French sociolinguist Henri Gobard (1976), denoting a field analysing

D. F. Virdis, *Ecological Stylistics*,
https://doi.org/10.1007/978-3-031-10658-3_2

cultural politics from an ethno-psycho-sociolinguistic perspective; afterwards, in an article by the Austrian-American behavioural psychologist Kurt Salzinger (1979), with a psycholinguistic meaning.

The term and discipline of ecolinguistics are founded on the noun "ecology". From a historical viewpoint, this noun was first employed by the German biologist Ernst Haeckel, one of Charles Darwin's followers, in a book printed in 1866 (Haeckel 1866, vol. 2, 286; see Sect. 6.1). Another main term in ecolinguistics is the noun "ecosystem"; it was coined by the English botanist Arthur George Tansley in a contribution appearing in 1935 in the journal *Ecology* (Tansley 1935; see Sect. 5.1). Derived from the noun "ecology", the adjective "ecological", in the 1960s, assumed a new denotation, which is now very frequent and has been recorded in the Oxford English Dictionary (OED Online 2022) as sense 2: "Concerned with environmental issues; environmentalist. Also: (of a thing or activity) environmentally friendly; causing minimal damage to the environment". This new denotation arose out of the ecological movement, which grew in those years and was strongly motivated by the American marine biologist Rachel Carson's (1962) book *Silent Spring* and by the American biologist Paul Ralph Ehrlich's (1968) volume *The Population Bomb* (Fill 2018, 1–2; Killingsworth and Palmer 2019, xix). Since then, the link between language and ecology has been explored in several ways and by means of numerous methods and approaches; since the early 1990s, ecolinguistics has grown and flourished internationally as a major area of research, teaching and theory development. This is demonstrated by the high and growing number of academic publications, conferences and workshops in the field around the world, and by the formation of the International Ecolinguistics Association (IEA), initiated and convened by the linguist and discourse analyst Arran Stibbe (for essential reference, see the References section at the end of this chapter; for complete reference, see Fill and Penz 2018a).

In their recent article about the state of the art of ecolinguistics, Steffensen and Fill (2014) made use of the conceptual metaphor ECOLINGUISTICS IS AN ARCHIPELAGO to refer to its diverse research interests and strands. Fill and Penz (2018a) identified three different but complementary concerns and lines of research in ecolinguistics during its forty-five-year history (for a previous classification in four strands, see

Fill and Mühlhäusler 2001). The first, chronologically, is also called "the ecology of language(s)": it deals with the interrelations between languages in their social environment (multilingual social system, country or continent) and individual environment (our human brain), and with their sustainability in the period of globalisation. It discusses topics like minority and majority languages; language revival, endangerment and death; bilingualism and multilingualism; contact languages. It also scrutinises the connection of linguistic and cultural diversity with biological diversity, and of language and culture loss with species loss; their large-scale extinction may be caused by the gross and increased inequalities between the rich and powerful peoples and countries and the poor and powerless ones, and the heavy weight of economic growth and rivalry (Fill and Penz 2018b, 438). Languages around the globe are being lost at a rapid and disconcerting rate; together with them, what is also being lost is the intimate experience of the natural world and the learning about the interaction between us humans and our environment developed and refined by generations over millennia; in sum, what is also being unrecoverably lost is sustainable ways of being on the planet, alternative hypotheses for knowing places and organisms, non-Western and non-Westernised mindsets and philosophies to access and interpret reality, and conceptual resources to deal with ecological issues (Harrison 2007, 3, 17).

This line of research evolved from the seminal work of the American linguist Einar Haugen (1972), who was the first to combine language with ecology. He wrote about the innovative idea of the "ecology of language", where "ecology" is a metaphor for the language system and the language world systems. The ecology metaphor is helpful for it throws light on a range of aspects of ecolinguistics and its subject matter: "the diversity of inhabitants of an ecology; the factors that sustain diversity; the housekeeping that is needed; the functional interrelationships between the inhabitants of an ecology" (Fill and Mühlhäusler 2001, 3). The ecology of language consequently examines the dynamic interrelations between a certain language and its social environment, viz. the society deploying it as a medium of communication: hence, its key priority is sociolinguistic. Two parts can be identified in the ecology of a language. Part of it is the connection between the language under investigation with other languages used by bilingual and multilingual speakers: accordingly, it is a psychological part. Another part of it is the connection between a language and the society in which it acts as a code: accordingly, it is a sociological part (Fill 2018, 3–4).

The ecological movement gave rise to the second line of ecolinguistic research, also referred to as "ecological linguistics". The subject of this strand is the part played by language, text and discourse in constructing our view of the natural world and in representing the environment in ecological and unecological ways. This line of research has two key interests. The first is the promotion of green grammar and criticism of the language system ("langue" in Saussurean terms). Its anthropocentric language structures depict the natural world from the perspective of its usefulness for us humans and our mainly commercial and economic activities, and predispose us to ecologically questionable actions and perceptions, like the objectification of living organisms; linguistic anthropocentrism should hence be replaced by an ecologically more correct linguistic physiocentrism and biocentrism.[1] The second interest is the analysis of discourse and language use (Saussure's "parole") and criticism of unecological ideologies and discursive options inciting the exploitation of animals and plants; research thus advocates good practices to create linguistic and ecological awareness of environmental problems (see the concept of ecoliteracy developed in the 1990s by Capra [1996] and Orr [1992]). This strand of ecolinguistics therefore explores matters like the following:

1. The function performed by language in addressing, euphemising, hiding and silencing ecological questions;
2. Ecological and unecological issues, representations and evaluations in a broad array of text-types by a broad array of social actors to achieve a broad array of objectives;
3. The practices deployed in language, text and discourse to express ecological and unecological considerations affecting the way we humans deal with the earth and creating unequal power relations with it;

4. Among recent developments, the discourses and ecological concerns of local communities and the standpoints of the people whose health, environment and democratic rights are damaged by unecological policies (Fill 2018, 3–5, Fill and Penz 2018b, 440).

This line of research was inspired by a talk and an article by the English linguist M.A.K. Halliday (Halliday 1990), where he agrees with the principle of linguistic determinism upheld by the American linguist Benjamin Lee Whorf. The world and nature are not passively mirrored by language, but are actively created and organised by language and the linguistic system in our minds; accordingly, diverse languages provide diverse perspectives on the world and nature. From an ecological standpoint, this means that language and grammar conceptualise natural materials (air, water, soil, etc.) as mass nouns, hence as unbounded materials, and us humans as disjointed from the other animal and vegetable species. Furthermore, language and its lexical and syntactic structures also support and disseminate the value-systems of speciesism, sexism, classism and, principally, growthism, or the expansion of anything human at the expense of damaging or even losing the natural world and the environment; as a result, growthism and the linguistic patterns indicating it have a direct impact on the existence and prosperity of our human life as well as more-than-human life on the planet (Fill 2018, 4–5) (for a different position from Halliday's, see Goatly 1996; see also the debate on green grammar between Andrew Goatly and Mary J. Schleppegrell reprinted in Fill and Mühlhäusler 2001, 203–231).[2]

A third line of research in ecolinguistics, emerging in the last two decades, scrutinises the philosophical side of human ecology. It considers how the field has started and established productive academic relations with other disciplines, such as philosophy, education, ethics and religion. The authors working within this strand conceive of ecolinguistics as an integrated, transdisciplinary and holistic ecological ideology; it signals the consciousness that our human world and the more-than-human world are interconnected and in harmony with each other. Consequently, in this line of research, ecolinguistics is not simply a linguistic area, but a multi-layered worldview. The latest development in this strand is the association with Chinese philosophy (Confucianism and Taoism);

here, ecolinguistics is held to be an ecological mindset and a mental and emotional condition highlighting harmony and interaction between us humans and nonhumans (Fill 2018, 5; for the new orientations and future directions of the discipline, see Fill and Penz 2018b; for eco-translatology, the application of ecological principles to Translation Studies, also originating in China, see Hu 2020).

While Fill and Penz (2018a) is the most authoritative handbook of ecolinguistics to date, Stibbe (2021), now in its second edition, is the most influential single-authored book-length study in this field so far, both theoretically and analytically; it is complemented by a companion website providing a free online course (Stibbe 2017). As previously shown, the three lines of ecolinguistic research presented above are based on the two primary denotations of the noun "ecology": a metaphorical denotation, underpinning the ecology of language(s), and a literal denotation, underpinning ecological linguistics. In Stibbe's (2021) view of the area, ecolinguistics principally falls into the second of the three strands. Hence, of the two denotations of "ecology", this scholar selected the literal one, which is equivalent to that offered by the Oxford English Dictionary (OED Online 2022) in sense 1.a (see Sect. 6.1): "The branch of biology that deals with the relationships between living organisms and their environment". Stibbe's book plans a new route for ecolinguistics as academically and critically engaged examination: the main goal of ecolinguistic investigation of text and discourse is to disclose what this scholar defines as the stories-we-live-by and to evaluate them against an ecosophy (see Sect. 1.3). As a result, ecolinguistics assesses positively beneficial discourses and patterns of language use encouraging us humans to support the protection of the natural world; it assesses negatively ambivalent and destructive discourses and patterns of language use impacting on ecological destruction. This field consequently assumes a central part in building a fairer, more equal and environmentally sustainable society. It is this strand of ecolinguistics and this denotation of the noun "ecology" that underlie the ecostylistic approach adopted in this book. Against this theoretical background, the following Sect. 2.1.1 defines discourse in ecolinguistic terms; Sects. 2.1.2–2.1.4 delineate what beneficial discourses are and present the aspects of ambivalent and destructive discourses.

2.1.1 Ecolinguistics and Discourse

In the research and teaching areas of discourse analysis and critical discourse analysis, discourse is, strictly speaking, spoken, written or sign language use or, broadly speaking, any meaningful semiotic or communicative event. Discourse is studied as a form of social activity or praxis, and expresses the various identities of the entities and individuals engaged in it and, most of all, their complex and multi-layered linguistic and non-linguistic dialectical relations. Fairclough (2010, 3) further asserts that.

> "discourse" might be seen as some sort of entity or "object", but it is itself a complex set of relations including relations of communication between people who talk, write and in other ways communicate with each other, but also, for example, describe relations between concrete communicative events (conversations, newspaper articles etc.) and more abstract and enduring complex discursive "objects" (with their own complex relations) like languages, discourses and genres. (Fairclough 2010, 3)

Discourse simplifies and condenses linguistically the articulate non-linguistic worlds of social and power relations. It portrays reality as it is or, to be more precise, as it is reckoned to be; more importantly, it is a linguistic practice systematically constructing the entities it denotes and the way we discern the world, conceptualises possible and imaginary realities distinct from the world as we know it, and is entangled with agendas to modify reality in line with specific trends (Fairclough 2010, 3–4).

In a rhetorical approach to the language of environment, discourses link ecological arguments and debates; environmental arguers wish to relay their values and authority and to make an impact on their addressees and culture. As a result, ecological discourse always concerns contrary opinions and beliefs, difference between people and divergence between the stances on nature and the earth that they adopt and that the media repeat and reproduce. Furthermore, ecological discourse often presents dichotomous languages and antithetical worldviews, a metaphor for which is "dispersed polarization" (Myerson and Rydin 1996, 204–206).

In the discipline of ecolinguistics, discourse is "The characteristic way that a particular group in society uses language, images and other forms of representation" (Stibbe 2021, 222). A given social group is recognisable by its distinctive use of language running across the multiple texts it produces, i.e. by such stylistic properties and patterns as grammatical selections and lexical choices. In this field as well as in ecostylistics, the investigation of a text does not concentrate on whether the discourse indicated by the text is accurate; what is appraised is whether the discourse urges its addressees to safeguard or devastate the ecological structures sustaining living organisms, viz. whether the discourse is consistent with the analyst's ecosophy. In addition, given that a discourse is a depiction, interpretation of or standpoint on reality outlined as true, but not necessarily based on fact, the addressees of a text proposing a specific discourse may not be aware that it is just a linguistic construct. Accordingly, the goals of ecostylistic and ecolinguistic scrutiny are, firstly, to reveal how discourse is encoded in language and bring the appropriate stylistic techniques to the fore; secondly, to advocate and champion discourses and techniques which are proven to be in line with the analyst's ecosophy, or to resist and dissent from those which are not.

2.1.2 Beneficial Discourses

Ecolinguistics has detected three dissimilar types of discourse dealing with ecological matters: beneficial, ambivalent and destructive. A beneficial discourse conveys the ideology of a beneficial story, which is defined as follows: "A story which accords with the ecosophy of the analyst (e.g. it is seen as encouraging people to protect the ecosystems that life depends on)" (Stibbe 2021, 222). In other words, if a discourse advocates such values as care for the planet, the wellbeing of the other animal and vegetable species on it, resilience, social justice and reduction in consumption, it is beneficial. Although it is not fully developed yet, looking for beneficial discourses and beneficial discursive practices is one of the core activities of ecolinguistics, consequently of ecostylistics (Stibbe 2021, 26–30). This is because discourses promoting more ecologically-aware attitudes are neither recurrent nor hegemonic in our nature-depleting and environment-degrading culture. Scholars of these areas should hence question the status quo by explicitly challenging our currently accepted and established view of the environment: they should support beneficial discourses and make them dominant as a cogent and preferable mode of life and system of knowledge and practice to those of the established social order.

Underlying beneficial discourses are intrinsic and larger-than-self values; as mentioned above and in the ecosophy of this research (norms no. 1: Valuing Living and no. 4: Care) (see Sect. 1.3), attention and concern are due to all human, animal and vegetable organisms. As maintained by Crompton (2010, 84), substantial data suggest that pro-ecological behaviours and stances are closely connected with and depend on intrinsic values; on the contrary, ecologically devastating behaviours and stances are related to extrinsic values, for example emphasis on individual advantage, material enrichment or economic position.

The idea of beneficial discourse and its underscoring of intrinsic and larger-than-self values draw on deep ecology, a social, political and spiritual philosophy fashioned around the environmental agenda, whose name was coined by the philosopher Arne Naess in the early 1970s. Deep ecology is a global ecocentric grassroots movement, in contrast to anthropocentric and technocratic theories, whose goal is to protect the planet for its own sake, to attain biocentric equality and ecological balance, and to make an "ambitious case […] for an extrahuman morality" (Harré et al. 1999, 183). It advocates a sense of reverence for the natural world and the "Rejection of the human-in-environment image in favor of *the relational, total-field image*" (Naess 1995a, 3). As a result, deep ecology depends on holism and on the idea that human reality and more-than-human reality fit together in a dynamic balance; consequently, there is no ontological divide between them. That is to say, deep ecology conceives of the planet as an organic whole composed of combined natural systems, and of us humans as merely one of a large number of components, and neither superior nor inferior to the others in the global ecosystem. All organisms in the biosphere have an equal intrinsic value independent of all human needs, and "the equal right to live and blossom" value axiom has relevance for all of them (Naess 1995a, 4; see also Naess 1995b). Deep ecology considers it a pressing and imperative necessity to simplify human life, to alter our culture and society radically and to implement ecologically responsible policies: this is in order for the earth's wellbeing to be renewed and sustained in the long run (Drengson and Inoue 1995). Furthermore, as opposed to what is designated as the shallow ecology movement (see Sect. 2.1.3), this philosophy takes a wide view of the term "ecology": in Naess' (1995a, 3) words, "There are deeper concerns which touch upon principles of diversity, complexity, autonomy, decentralization, symbiosis, egalitarianism, and classlessness".

The first ecostylistician and one of the first ecolinguists to identify and investigate a beneficial discourse was Goatly (2000, 285–301), who compared the beneficial portrayal of nature made in Romantic poetry with its destructive representation in newspapers. In the latter, the natural world is stylistically shaped as comparatively inert and tamed, borne in mind merely with regard to weather, disease, horses and dogs we humans are fond of, and subjugated to economic worries. By contrast, in the former, especially in Wordsworth's poems, the physical environment is fashioned by the grammar as teeming with life and far more dynamic. The poems therefore reappraise our interaction and interconnections with nature with a view to changing our attitude to it; they offer a model ideally suited to the ecosophy of this research and its major goal of living (norms no. 1: Valuing Living and no. 2: Wellbeing) or, worse, surviving on a planet which will probably grow more and more hostile to life (norm no. 7: Adaptation) (for Romantic nature writing, see Bate 1991; Coupe 2000).

Stibbe (2021, 27–29) supplies several past and present instances of nature writing from all over the world constituting beneficial discourses and models and, as such, requiring further ecolinguistic or ecostylistic scrutiny (for a recent overview of the genre of nature writing, see Hummelsund Voie 2019). Among them are the works of such groups as Imaginative Naturalists and contemporary British New Nature Writing (both termed so by Macfarlane [2013]); the lifestyles evoked by Japanese *haiku* nature poetry and Native American literature; concepts bequeathed by native and traditional cultures, like the Japanese *satoyama* (literally meaning "homeplace/mountain" and denoting the border zone between mountain foothills and flat arable land where nature and culture are interconnected) (for the philosophy of *satoyama* and traditional Japanese ecological knowledge, see Masami 2019).

After beneficial discourses have been identified and their language and stylistic items explored, they should be advertised. Advertising a discourse or promoting stylistic items is not tantamount to advertising or promoting certain texts; it means advertising the discursive mechanisms and the clusters of discursive mechanisms conveying a helpful worldview and discourse in accordance with the analyst's ecosophy. This implies that discourses and discursive mechanisms can cross texts and text-types; as a result, those thought to be beneficial can be adjusted and included throughout an extensive variety of text-types employed in and characterising various spheres of knowledge. The final thrust of advertising beneficial discourses is to include them readily in our prevailing social, economic and cultural discourses, so that they can be a significant influence on our everyday life and activities (Stibbe 2021, 29–30). These are the principal reasons why this book analyses diverse websites from the environet (Myerson and Rydin 1996, 7; see Sect. 1.1), and detects and promotes their ecostylistic strategies relaying beneficial discourses.

In his book about discourses on animals and our different conceptualisations and descriptions of them, Stibbe (2012, 121–143) utilises the terms "alternative" and "counter" discourses rather than "beneficial" and "ambivalent" discourses, respectively; the term "destructive" discourse remains unchanged (for the discursive representation of animals from an ecolinguistic perspective, see Cook and Sealey 2018). An alternative discourse is founded on a belief-system dramatically dissimilar to the established belief-systems underpinning counter and destructive discourses and modern unsustainable cultures, in that an alternative discourse urges us humans to reconnect with nature in general and animals in particular. This researcher (Stibbe 2012, 123–135) examined the ecological discourse employed in a sample of English-language textbooks written by native speakers and distributed in Japanese universities. Although their overall environmental discourse is based on shallow ecology (see Sect. 2.1.3), consequently it is a counter-discourse, the textbooks feature a small number of sequences communicating an alternative discourse and deep ecology considerations. For example, sympathy for other animal and vegetable species experiencing pain and death due to ecological ruin is advocated by stylistic choices like utilising blunt and non-euphemistic terms, specifying the suffering animals by name and referring to their number to emphasise their individuality, accordingly stirring the readers to compassion (Stibbe 2012, 137; for animal euphemisms, see Trampe 2018). Another example of alternative discourse derives from traditional Japanese culture: the recitation of the word *itadakimasu*, which means "receive". The Japanese observe this ritual with ecological associations when sitting down to a meal; the action of reciting the word is an expression of gratitude and acknowledgement to the animals and plants whose lives were taken for the meal. *Itadakimasu* is a component of *naikan*, a social and cultural value-system championing the grateful and appreciative use of the smallest possible amount of natural materials, hence in contrast to consumption and wastefulness (Stibbe 2012, 138).

2.1.3 Ambivalent Discourses

Moving from beneficial discourses to ambivalent ones, a discourse is called ambivalent when it conveys the ideology of an ambivalent story, or "A story which only partially accords with the ecosophy of the analyst (e.g. it is seen as having mixed benefits and drawbacks in encouraging people to protect the ecosystems that life depends on)" (Stibbe 2021, 221). If a discourse has the positive objective of sorting out ecological issues but springs from and is embedded in the same culture, society and economy as these issues, and depends on financial or political tenets, it is ambivalent: it commonly appears to be a green discourse, but it actually preserves the status quo. A number of ambivalent discourses of environmentalism conceive of the natural world not as an end in itself, but as a means to be preserved exclusively in order for us humans to take advantage of it indefinitely. Ambivalent discourses are generally frequent and prevailing, so much so that they are the most explored in ecolinguistic analysis. Since many of the objectives of beneficial ecosophies and of the ecosophy of this research are often found in ambivalent discourses, these discourses should not be opposed as a whole: their questionable features should be tackled and their constructive features preserved and promoted (Stibbe 2021, 25–26).

While the concept of beneficial discourse derives from the philosophy of the deep ecology movement, the notion of ambivalent discourse originates from the shallow ecology movement. In Naess' (1995a, 3) words, the latter can be characterised and summarised as follows: "Fight against pollution and resource depletion. Central objective: the health and affluence of people in the developed countries". Shallow ecology, which is still the prevailing trend of ecological opinion among the general public, can also be called corporate or technological environmentalism (Drengson and Inoue 1995, xix) on the following grounds:

> It advocates continuous economic growth and environmental protection by means of technological innovation (such as catalytic converters), "scientific" resource management (such as sustained yield forestry), and mild changes in lifestyle (such as recycling). It avoids serious fundamental questions about our values and worldviews; it does not examine our sociocultural institutions and our personal lifestyles. (Drengson and Inoue 1995, xix)

That is to say, the fundamental elements of shallow ecology are its being moderately reformist in nature and its anthropocentric prejudice that animals and plants only serve as instruments to achieve our human ends and purposes, mostly in the advanced industrialised states (Drengson 1995, 76). Moreover, contrary to deep ecology, shallow ecology does not treat the underlying causes of the environmental crisis, namely our worldviews and consumerist societies and values; it just treats its physical symptoms by means of surface changes and technological fixes like the consumption of green or environment-friendly products, regulations for industry and pollution-control devices (Rothenberg 1995, 155). This recalls Myerson and Rydin's (1996, 208) concept of answer culture: "The 'answer culture' is a metaphor for certain expectations: that knowledge will save us".

Considering the contradictory and problematic character of shallow ecology, some parts of ambivalent discourses incline to more beneficial values, while others lean towards more destructive priorities. A case in point is the ecological discourse and communicative strategies of such companies as Marks & Spencer, Tesco and The Body Shop. Although all their websites feature pages about beneficial topics like sustainability, fair trade and food waste, their discourses and strategies are explicitly ambivalent (if not downright destructive) by their very nature. They commonly appear to be green discourses but, since they promote commercial companies, they cannot but preserve the status quo and support consumption. These discourses are instances of greenwashing, or misleading publicity designed to present a corporate image of ecological responsibility (Ihlen 2009; Howlett and Raglon 1992; Stöckl and Molnar 2018). Greenwashing is based on Greenspeaking (Harré et al. 1999; Mühlhäusler and Peace 2006), the "greening" of language which also encompasses discursive strategies verbally replacing or postponing concrete positive action with abstract ecological words, expressions and discourses. These words and expressions were first introduced into English, other Western languages and their political, institutional, economic and moral discourses some fifty years ago; since then, they have increasingly been absorbed into everyday language and the dominant discourse of Western culture; as such, they have lost their illocutionary force, a phenomenon characterised as "semantic bleaching" (Mühlhäusler 2000).

Another case in point is the ambivalent ecological discourse and communicative practices in a key official document about the environment and sustainable development, issued by the United Nations General Assembly (UNGA): the Rio Declaration on Environment and Development (UNGA 1992). This is the achievement of the United Nations Conference on Environment and Development (UNCED) held in Rio de Janeiro, Brazil, on 3–14 June 1992. The Rio Declaration is unanimously considered to be an impressive landmark in the advancement of the ecological legal system at a global level, and to have drawn a boundary line between the old age and the new age of international law on the environment (Handl 2012). The Rio principles and action plans were reaffirmed in the Rio + 20 Outcome Document, also known as The

Future We Want (UNGA 2012), and in the 2030 Agenda for Sustainable Development (UNGA 2015). Along with these two documents, the Rio Declaration energetically promotes sustainable development, which intermingles ecological conservation and social justice with economic growth (Baker 2006). Nevertheless, economic growth is supported not only in developing nations but also in states which are already industrialised and overdeveloped, and is commonly in the foreground, while ecological and social principles are normally in the background.

As anticipated in Sect. 2.1.2, when animals and animal representations are the focus of a text or a sequence from a text, the term "counter" discourse is preferred to "ambivalent" discourse (Stibbe 2012, 63–83). A counter-discourse fails to escape from the restraints of a destructive discourse. For instance, the English-language textbooks aimed at Japanese university students discussed above promote harmony with nature and denounce ecological destruction, but at the same time they advertise anti-ecological assumptions. Among them is the assumption that more-than-human life has no intrinsic worth (Stibbe 2012, 129–131). This devaluation is expressed by animal and vegetable organisms metaphorically portrayed as resources, assets and raw materials, or as collectables to be deposited and seemingly extolled in museums and zoos for educational reasons. Devaluation of more-than-human life in the textbooks is also stylistically realised by representing ecological devastation as us humans suffering financial loss from the depletion of commercially valuable resources, rather than as huge numbers of animals and plants suffering agonising pain or death. A further stylistic technique to devalue more-than-human life is identifying animal and vegetable organisms as species in their entirety only; in other words, living organisms are not conceptualised as independent or separate animals and plants to be protected as such. The textbooks do not denounce the ruinous courses of action aimed at nonhuman individuals or communities whose deaths do not openly affect the existence of the species they belong to; these books exclusively disapprove of the devastation of whole species and describe their endangerment and extinction as commercial, not ecological, tragedies.

2.1.4 Destructive Discourses

When a discourse is destructive, it conveys "A story which opposes or contradicts the ecosophy of the analyst (e.g. it is seen as encouraging people to destroy the ecosystems that life depends on)" (Stibbe 2021, 222). Namely, if a discourse contributes to ecological exploitation and devastation, as well as to social oppression, and if it continues to ignore our human mistakes resulting in the present environmental crisis, it is destructive: the ideologies it conveys are usually disseminated extensively and we should be empowered to resist them. Like ambivalent discourses, destructive discourses are recurrent and influential, which is why they are normally selected for critical ecolinguistic examination: as Harré et al. (1999, 137) put it, "Greenspeaking [...] characteristically contains a considerable amount of self-criticism".

The most recurrent and influential destructive discourse is possibly the discourse of economics. As claimed by Gare (1996, 132–144), in our contemporary world economics is the discourse which, for most people, establishes and interprets the character, essence and scope of our human existence and society: it furnishes the ideas to delineate and justify our human connections with our fellow humans, their social system and nature. The prevailing discourse of economics does not often refer to the environment directly; however, it lays down associations between us humans and the planet which are commonly estranged, therefore destructive (see also the pervasive discourse of the market society or marketisation, which has been proven to be one of the hegemonic worldviews underpinning our social systems [Lakoff 2010; Mautner 2010]).

For these reasons, since the early days of ecolinguistics, the discourse of economics has been a challenge for the field and has provided linguistic data for investigation. Halliday (1990) studies how economic growth is often conceptualised by language as one of the ultimate purposes of our world and culture; these conceptualisations are to blame, as unlimited growth on a limited earth will bring about a drain on nature and the devastation of the ecological structures sustaining life (see Sect. 2.1). Language inciting growth is also accused by Goatly (2000, 278): in the case of what he labels "mature economies", he recommends that the noun "growth" be replaced with the metaphorical noun "cancer", in order to underline that nature and its ecosystems are endangered by further growth in an economy which is already mature. Chawla (1991) notes that the sound condition of our human existence is quantified in terms of a continuous expansion in the production of saleable commodities and in terms of constantly increasing levels of consumption (for the term "growth", see Harré et al. 1999, 29–31). This contradicts the ecosophy of this research for at least two reasons: (1) it advocates needless overproduction and overconsumption, which trigger ecological harm (against norms no. 5: Environmental Limits and no. 6: Social Justice); (2) it conceals all the means to physical and psychological health not entailing the consumption of goods (against norm no. 2: Wellbeing).

The economic discourse of overproduction and overconsumption is linked to and incited by advertising discourse which, according to Chawla (1991), changes us into consumers and makes us feel unfulfilled with any standard of consumption we may obtain. The discourse of advertising is one of the most ecologically devastating discourses, since its primary functions are to convince us to purchase commodities which had not been regarded as essential before and to persuade us that those commodities are a quick way to achieve wellbeing: one key ideology that advertisements disseminate is, in Stibbe's (2021, 23) conceptual metaphor, PURCHASING PRODUCTS IS A PATH TO HAPPINESS. One of the major strategies deployed in the discourse of advertising consists of the assumption that buying goods can suitably and glamorously replace desirable activities and pursuits which would bring deep pleasure at no cost, such as dancing, roaming the countryside and taking exercise. The discourse of advertising is confirmed and reinforced by the discourses of lifestyle and consumer magazines and by the discourses of aspirational television programmes. They deliberately fabricate discontent, thereby inciting us to overcome it by purchasing the goods advertised in those magazines and programmes and by defining our identities by means of those goods. Fabricating discontent acts counter to wellbeing and urges us to overconsume; as a result, from an ecological and ecostylistic viewpoint, this damages the environmental structures life depends on.

Influential destructive discourses like the discourse of economics and the discourse of advertising should be resisted. This means increasing our consciousness that these discourses suggest ideologies which, in reality, are merely stories and prevailing assumptions sustaining power inequalities (Stibbe 2012, 21–22), and that they affect adversely the natural world we humans belong to. Consciousness-raising also applies to the people in charge of employing and propagating destructive discourses, such as economists, politicians, advertisers and journalists. Indeed, one of the principal aims of ecostylistics and ecolinguistics is to inform them about the mindsets they are championing and their harmful ecological influences, principally if they are doing so unintentionally. Consciousness-raising also has relevance for the addressees of destructive discourses, voters and customers in particular, who should be strongly urged to make more informed choices, especially when the addressers of these discourses are reluctant to alter their stylistic practices. The main goal of resisting destructive discourses is to reduce their widespread use and to promote more beneficial options (Stibbe 2021, 24).

The term "destructive" discourse is also utilised to refer to texts or sequences centred on animals and animal depictions (Stibbe 2012, 19–33). When applied to these texts and sequences, the term refers to discourses and linguistic practices sustaining the inhumane treatment of animals along with ecological harm. In general discourse (words and phrases not related to specific categories of users, but deployed in everyday interaction in a broad array of discourses [Stibbe 2012, 23]), a destructive use of language lexicalises and ideologically positions animals as different from and inferior to us humans through negative, if not cruel, collocations, metaphors and idioms (Heuberger 2018). The most revealing example is perhaps the semantics of the very word "animal". In scientific discourse, the noun and adjective "animal" denotes human animals and nonhuman animals alike; in contrast, in general discourse, it only denotes nonhuman animals. The semantics of the word in general discourse psychologically dissociates us human animals from nonhuman animals and their suffering, thus ultimately resulting in their oppression.

A destructive discourse where animals and their individuality are most clearly misrepresented is the discourse of agribusiness and the animal product industries (Stibbe 2012, 25–31). It is founded on the ideological supposition that animals are inanimate resources unable to suffer, which leads to their abuse, oppression and exploitation in these industries.

Abuse, oppression and exploitation are justified and described as natural and inevitable by means of the apparent objectivity of the discourse of science and evolutionary biology. The cruel and merciless world of intensive factory farming and slaughter is normally disguised as preoccupation with animal welfare and falsely depicted through lexical choices. For instance, farmers are often portrayed as their animals' protectors; the noun "cage" is seldom employed and generally replaced by euphemisms like "housing"; metonymy turns "chickens" into "broilers" and "female animals" into "breeders" (Trampe 2018).

In destructive discourses, animals are often named collectively: they are referred to through mass hypernyms like "wildlife", "marine life", "fauna" and "by-catch", and their quantity is determined in terms of weight (e.g. "tons") and not in terms of the number of the individual animals. These stylistic techniques obliterate the fact that animals are typified by attributes distinguishing one from the other; underpinning these techniques is the presumption that every animal is a mere specimen of their species and, as such, is interchangeable with any other. In post-modern philosophical words, this is the mental mechanism of erasure (Olson 2000), "a practice that may well be a precursor to their [animals'] physical erasure" (Stibbe 2012, 131).

2.2 Stylistics

Along with ecolinguistics, the other discipline inspiring ecostylistics as conceived of in this volume is stylistics. Stylistics is the study and analysis of texts founded on precise and exhaustive linguistic description, and is characterised by a close reading and explanation of the texts under investigation. In addition, the discipline is also characterised by the fact that the analysis of how interpretation is carried out by a stylistician should be easily replicable by their fellow stylisticians and by readers (Simpson 2014, 4). The research focus of stylistics is, accordingly, the linguistic makeup of texts and the centrality of language in textual construction. The area has developed a theoretical framework and a wide variety of linguistic and analytical strategies capable of critically, scientifically and systematically "interrogating" not only language, texts and textual meaning, but also readers and contexts. That is to say, stylistics can expose, question and challenge not only linguistic meaning-making practices and the textual features readers respond to, but also cultural and social ideologies, viewpoints and belief-systems underlying individual texts and their social and cultural contexts. Although a unified, coherent and specialised discipline since at least the 1990s, contemporary stylistics is a heterogeneous field applying diverse methods. It is designed to be interdisciplinary and eclectic in the theories it embraces and the methodologies it adopts and adapts; to name just a few, they range from pragmatics, pedagogy, cognitive linguistics and gender studies to corpus linguistics, historical linguistics, multimodality and even neuroscience (Burke 2014; Jeffries and McIntyre 2010; Sotirova 2016; Stockwell and Whiteley 2014).

With regard to the stylistic methodology utilised in this book, "stylistic explanation of the relations between textual form and literary function is a specialist grammatical commentary. Accordingly, a theory of literary stylistics is at core a theory of grammar" (Toolan 2014, 20). Toolan considers the term "grammar" in its widest sense, ranging from the description of word, phrasal and clausal sequencing to genre and register conventions and practices. This seems to be in line with Quirk et al.'s (1985, 11–12; 14–15) approach to grammar. In their view, grammar studies the articulate set of norms applying to the combination of morphemes into words (morphology) and to the combination of words into larger units (syntax). Nevertheless, the study of words, their form and their meaning is also the concern of lexicology, while the study of meaning relations in the language system is the object of semantics: both of these linguistic branches are consequently relevant within grammar and interrelated with it. Given these scholars' definitions, the grammatical, hence stylistic, levels variously scrutinised in this book are syntax, lexis and semantics.

More precisely, the methodology adopted here to examine such grammatical levels is defined by Carter (2010, 58–61) as "practical stylistics" or "steam stylistics"—the latter term used, in a jocular way but in the most positive sense, in contrast to recent developments in corpus stylistics and cognitive stylistics (see also Gregoriou 2014, 95). Practical stylistics is founded on the assumptions that a text is made from language and with language, and that its textuality and linguistic makeup are key to its interpretation. Drawing from practical criticism, practical stylistics is based on the analyst, their intuition, observation of the text and knowledge of the language. In fact, it entails close reading of the linguistic texture of the work under investigation and the systematic identification of the significant choices and patterns in it; special attention is paid to the characteristics of language and textual organisation recognised as foregrounded or deviating when contrasted with more standard uses. A practical stylistic study is undertaken according to an organised method; this is also with a view to the study being retrievable and replicable by other analysts, as all stylistic examination must be. This organised method can involve checklists, as put forward by Short (1996) and Leech and Short (2007, 60–66); for this reason, three checklists to explore the stylistic strategies of foregrounding, point of view and metaphor are offered in the Appendix to this book (Chapter 9). The primary aim of practical stylistics is to link linguistic form and textual meaning: namely, to explain the processes of meaning construction, to show how linguistic choices give rise to interpretations and to account for the effects such choices produce. The key stylistic devices found in the texts scrutinised in this book and emerging through the application of practical stylistics are foregrounding, point of view and metaphor. They are defined and treated in the three sections below.

2.2.1 Foregrounding

The literary-linguistic theory of foregrounding is connected with the notion of defamiliarisation. This notion holds that an addressee's attention and curiosity are aroused by texts portraying ordinary people, activities and events in an unordinary, fresh and emphatic manner (Šklovsky 2012). This process can be favoured by specific linguistic traits awakening the addressee's interest and conveying defamiliarisation: this is the theory of foregrounding (Mukařovský 1964). Foregrounding can be accomplished through a wide array of stylistic techniques. The list of foregrounding devices compiled by Emmott and Alexander (2016, 290) encompasses such linguistic tools as "repetition, odd punctuation, under-specification and over-specification, interesting vocabulary and grammatical choices, striking similes and metaphors, recurrent thematic motifs, unusual narratological presentation and unusual textual layouts".

To be more precise, the Russian Formalist term *ostranenie*, first deployed in 1917 by Victor Šklovsky and usually translated as "defamiliarisation", indicates an artistic estrangement from what are regarded as everyday entities and customary facts, which are thereby reappraised. The goals of this practice are to increase the addressee's attention to and engagement with a text, to intensify the depth of processing and the effort they put into perceiving and experiencing the text, and to extend the duration of that perception and that experience; moreover, defamiliarisation raises the addressee's consciousness that the process of perception is occurring in their mind. Due to the complexity of the process, a text spurring conscious perception is assumed to be a form of art, and deliberate defamiliarisation is conceived of as a marker of aesthetic effect and cultural value (Emmott and Alexander 2016, 291; see also Gavins 2014).

Defamiliarisation is attained by means of the linguistic and textual practice called foregrounding. First utilised by Jan Mukařovský in 1964, this Prague School term is openly metaphorical and based on the terminology of painting. The linguistic and textual realisation of defamiliarisation is likened to a perceptually prominent element placed in the foreground of a picture, i.e. in front and nearest to the observer; the standard parts of a text are compared with less noticeable elements falling into the background (see also the opposite technique of burying [Emmott and Alexander 2014, 331–332]).

As a result, the stylistic process of foregrounding designates words, phrases and clauses written in non-standard language; they are used by an addresser to catch their addressee's interest, to direct and manipulate their attention and, consequently, to encourage them to search for unstated meaning. Foregrounding infringes implicit linguistic assumptions and schemata by interrupting the normal continuity of the usual mechanical unmarked language and by making it unusual, non-mechanical and marked: old entities and realities are phrased and, thus, viewed in a fresh light (Emmott and Alexander 2016, 291). Foregrounding can have various functions; among them are those detected by Emmott and Alexander (2014, 329), i.e. "highlighting specific key points, producing thematic meaning, prompting an emotional response, and yielding iconic effects". Like defamiliarisation, foregrounding is reckoned to mark artistic effect and literary interpretation; foregrounding patterning, however, can also take place in everyday language and non-literary discourses with a rhetorical and aesthetic thrust. Hence, this book scrutinising foregrounding in non-literary texts follows Emmott and Alexander's (2014, 330) use of this term:

> We use the term "foregrounding" here to cover any type of language use which may be assumed to prompt attention, regardless of whether it has literary value. Foregrounding in this sense relates simply to whether an item is likely to be noticeable or not, and our interest is in whether it has rhetorical significance generally rather than literary significance specifically. (Emmott and Alexander 2014, 330)

As contended by Mukařovský (1977), who explored literary texts, one of the main foregrounding mechanisms is repetition, which he found to be particularly recurrent in poetry (see, for instance, the repetition of the verb in "Blow, blow, thou winter wind" [Shakespeare, *As You Like It*, II, vii, 174]). According to Leech and Short (2007, 39–41), who belong to the successive generations of stylisticians, repetition should be included in the more extensive foregrounding strategy of stylistic deviation. This idea implies stylistic distortion and unpredictable divergence from a linguistic rule or statistical norm, and consists of irregularity or over-regularity (parallelism) against a regular background. All practices of foregrounding constitute aesthetically-motivated and meaning-making deviation; hence, this is commonly associated with unordinary ways of producing text and discourse which may be considered favourably from a cultural, social and rhetorical perspective, like literary texts. Short (1996, 36–59) underscores the fact that deviation may be present at any linguistic or narrative level; accordingly, the addressee's curiosity may be aroused by salience and prominence at all levels of the language. This stylistician distinguishes different kinds of deviation:

1. Discoursal deviation. Short (1996, 37–43) supplies a number of examples of this type of deviation, among which are: (1) James Joyce's *Finnegans Wake* not beginning at the beginning of a sentence, but starting and ending in the middle of the very same sentence; (2) John Donne's "The Canonization" and Edward Thomas' "Adlestrop" not starting at the beginning of a discourse, but beginning *in medias res* and projecting the addressee into the middle of a conversation; (3) William Butler Yeats' "Long-Legged Fly" not featuring one prototypical addresser (the poet) and one prototypical addressee (their reader), but three different addressers and three different addressees.
2. Semantic deviation. This can be defined as "meaning relations which are logically inconsistent or paradoxical in some way" (Short 1996, 43). Consequently, semantic deviation can be realised by metaphors and paradoxes, for instance, Shakespeare's "Come, we burn daylight, ho!" (*Romeo and Juliet*, I, iv, 43) and Dylan Thomas' "Light Breaks Where No Sun Shines" (which is the title of a poem, its opening line and the first of a series of paradoxes in the text).
3. Lexical deviation. In Short's (1996, 45) words, "The most obvious examples of lexical deviation are those where a poet makes up a word which did not previously exist. This is usually called neologism". Instances of this kind of deviation are the noun "manunkind", coined by e.e. cummings ("pity this busy monster, manunkind"), and the adjective "jollyrodgered", created by Dylan Thomas (*Under Milk Wood*) by adhering to the technique of functional conversion. Short (1996, 46) continues: "Another way in which writers can produce foregrounding through lexical deviation is by taking words associated with one variety of English and inserting them in a variety where they do not normally belong". This mechanism is applied in Philip Larkin's poem "Send No Money", where the coarse slang expression "Sod all" is employed in a formal context.
4. Grammatical deviation. The most recurrent case of this type of deviation, very frequent in poetry, is realised by word re-orderings inside the noun phrase, with the adjective following the noun, or inside the clause, with the subject following the predicator: see, for example, "remembrance dear" (John Keats, "Ode to Psyche") and

"Then spake King Arthur to Sir Bedivere" (Alfred Lord Tennyson, "Morte d'Arthur") (Short 1996, 47–48).

5. Morphological deviation. As asserted by Short (1996, 51). "One way of producing deviation at a morphological level is by adding an ending to a word it would not normally be added to"; this is the case with e.e. cummings' "perhapsless" ("from spiralling ecstatically this") and James Joyce's "museyroom" (*Finnegans Wake*). Short (1996, 52) adds: "Another way in which poets can make us contemplate the otherwise unremarked morphological structure of words is by playing around with word boundaries"; an instance of this strategy is the noun "king- / dom", written on two distinct lines (Gerard Manley Hopkins, "The Windhover"). Furthermore, "The opposite process to separating morphemes off as quasi-separate words is to run more than one word together as if they were one"; this practice is exemplified by the word "Bothatonce" (e.e. cummings, "she being Brand").
6. Phonological and graphological deviation. According to Short (1996, 54), "because literature is written, the written equivalent of the phonological level can be much more liberally and usefully exploited"; this written equivalent is the graphological level. In Ted Hughes' "Crow's First Lesson", four occurrences of the word "love" are deviant: three of them are capitalised to suggest that the word has a prominent pronunciation, one occurrence is written entirely in capitals to evoke an even more underscored delivery of the word. Hence, Short (1996, 55) claims that "The repetition and the increasingly deviant *graphological* forms lead us to an interpretation with an increasingly marked *phonetic* form". Nevertheless, graphological deviation does not necessarily entail phonological deviation. In Laurence Sterne's *Tristram Shandy*, two black pages convey mourning for parson Yorick; the noun phrase "the next marbled page" does not designate a page of prose in grand and impressive style, as could be expected, but a page coloured with a variegated pattern; a long series of asterisks relays whispering at great length about socially delicate topics (for further non-literary examples of deviation than those furnished by Short, see Gregoriou 2014, 90–94).

While deviation produces foregrounding through irregularity, parallelism does so by means of over-regularity. Parallelism is linguistically realised when "some features are held constant (usually structural features) while others (usually lexical items, *e.g.* words, idioms) are varied" (Short 1996, 14). A case of parallel structure is *Isaiah*'s (53, v) "But he was wounded for our transgressions, he was bruised for our iniquities". The "'parallelism rule' for interpretation" applies: these features "invite the reader to search for meaning connections between the parallel structures, in particular in terms of the parts which are varied" (Short 1996, 14; see also 15–16, 63–68, and Gregoriou 2014, 94–95).

A further development in the examination of defamiliarisation, foregrounding and deviation can be found in the fields of cognitive stylistics and of the empirical study of literature, and revolves around the psychological process of reader perception. An important empirical analysis was undertaken by van Peer (1986) on readers of poetry, in order to demonstrate that foregrounding is a concept the addressee of a text deploys when placing an interpretation on it. All the cognitive contributions to this area have furnished psychological data on the existence of foregrounding and on the influence of foregrounding means on the activity of reading and explaining texts.

In this book, in the five analytical Chapters 3–7, foregrounding is investigated in line with the model provided by Leech and Short (2007, 39–41), and in line with the classification of the kinds of deviant linguistic usage and micro-level stylistic choices offered by Short (1996, 36–59), which is the most established sphere of foregrounding studies. The notion of foregrounding is therefore employed to identify occurrences of deviation, to supply a description of their salient linguistic aspects and to account for them as meaningful and pertinent to the overall message relayed by the text.

2.2.2 Point of View

One of the most thoroughly scrutinised fields of stylistic research, and a central interest in literary criticism since the birth of the modern novel in the eighteenth century (Neary 2014, 175), the concept of point of view has evolved to explore narrative literary texts (to name just one book-length study, see Simpson 1993). However, it is such an effective detailed analytical tool that it can also be successfully utilised to examine non-narrative and non-literary texts, like those studied in this book; accordingly, this section defines this concept and outlines its aspects, highlighting those having more relevance for all text-types only. The figurative umbrella term "point of view" refers to the physical, psychological and ideological angle a text is written or spoken from, i.e. to the writer's or speaker's perspective of an action, event, object or thought. A distinction must be made between what the text depicts and the standpoint it is depicted from, namely between who tells and who sees; this distinction in point of view is a factor bringing to light the many-sidedness of any text. Point of view is marked and indicated by an articulate interlace, interaction and patterning of linguistic items, which in turn mark and indicate distinct angles and voices. In Simpson's (2010, 295) words,

> An account of these traditional markers will include (but is not restricted to) categories such as the following: *deixis*, which is a set of "pointing" or "orientational" words ("this" vs "that", "come" vs "go") that function to attach a speaker/narrator to a particular context (and see further Short 1996, 269-272); *locative expressions*, which are phrases indicating place, location, and direction ("under the tree", "around the corner", and so on), and which anchor the spatial position of the reflector of fiction (and see further Fowler 1986, 128-130); and *verba sentiendi*, which are words [...] which serve to represent thoughts, feelings, and perceptions of characters or narrators (and see further Uspensky 1973 *passim*). (Simpson 2010, 295)

The most authoritative model of point of view is likely to be that first put forward by the narratologist Boris Uspensky (1973), which was subsequently evolved by the stylistician Roger Fowler (1996, 160–184). This model provides a taxonomy of the distinct types of point of view organised into four parts or planes: spatial, temporal, psychological and ideological. Spatial point of view is the phenomenon in its literal and non-figurative meaning: it designates the location or visual angle a narrator tells a story and relates facts from. Temporal point of view involves how the narrator perceives and conveys time, that is to say, whether events appear to occur at a fast or slow pace, and with or without breaks in continuity; for example, flashbacks (analepsis), flash-forwards (prolepsis) and narrative gaps are all mechanisms of temporal viewpoint. As anticipated in the Simpson citation above, linguistic indicators belonging to the system of deixis are usually among the key markers of spatial and temporal point of view alike. Psychological point of view denotes how a text, with the representation of the actions and objects in it, can be influenced by the distinctive perception, awareness or understanding of a narrator or character; psychological point of view and the characteristic subjectivity it communicates are frequently expressed through speech and thought presentation (see Note 3) and through *verba sentiendi* (see the Simpson quotation above). Ideological point of view indicates how a text relays specific value-, belief- and knowledge-systems by means of its author, narrator or characters; among its major indicators is language choice, or stylistic and lexical selections (Simpson 1993, 11–19, 27–39, 96–101; summarised in Neary 2014, 177–178).

According to Simpson (2010, 295), "Other traditional markers of point of view include *modality* which is that part of the system of language which allows speakers to attach expressions of belief, attitude and obligation to what they say or narrate". The four modal systems of the English language are deontic, boulomaic, epistemic and perception (Simpson 1993, 43–47; summarised in Neary 2014, 181–182). The deontic modal system embraces those words and structures signalling an addresser's stance on the degree of duty or commitment (permission, obligation and requirement) assigned to the performance of given actions; therefore, this system is intrinsic to the communicative and social practices of politeness and persuasion. The deontic system is indicated by:

1. Modal auxiliary verbs (*must*, *should*, *may*);
2. Adjective and participle patterns like "be + deontic adjective + that" and "be + deontic participle + to" (*It is essential that she do it*; *She is allowed to do it*).

The boulomaic modal system concerns the communication of an addresser's desires and wishes. These are suggested by the following words and structures:

1. Modal lexical verbs (*hope*, *wish*, *regret*);
2. Modal adverbs (*Hopefully, she will do it*);
3. Such adjective and participle patterns as "be + boulomaic adjective + that" and "be + boulomaic participle + to" (*It is good that she is doing it*; *It is hoped that she will do it*).

The epistemic modal system regards an addresser feeling more or less certain of or committed to the veracity and factuality of a specific proposition, and represents the linguistic concepts of knowledge, belief and cognition; as a result, it is perhaps the most notable modal system as far as the analysis of point of view is concerned. Epistemic modality can be relayed by certain words and structures:

1. Modal auxiliary verbs (*She could be wrong*);
2. Modal lexical verbs (*She supposes she is right*);
3. Modal adverbs (*maybe*, *possibly*, *perhaps*, *probably*, *certainly*);
4. Adjective patterns like "be + epistemic adjective + that" and "be + epistemic adjective + to" (*It is certain that she is wrong*; *She is certain to be wrong*).

The perception modal system consists of a sub-system of epistemic modality, since the degree of an addresser's confidence in the truth of a proposition is founded on visual and other, commonly human, powers of perception. Perception modality is communicated by:

1. Modal adverbs (*Obviously, she is wrong*);
2. The adjective pattern "be + perception adjective + that" (*It is obvious that she is wrong*).

A modal grammar of point of view was developed by Simpson (1993) on the assumption that the notion of modality and the four English modal systems can be deployed to methodically recognise the linguistic tools evoking diverse points of view. Distinctive frameworks of modality incarnate distinctive points of view and, when frequent, delineate the distinctive styles of given text-types. For instance, modal patterns involving epistemic modality, viz. knowledge and understanding, are emphasised in the horror fiction genre and bring about a characteristic type of narrative perspective communicating the narrator's efforts to explain and figure out the world surrounding them. In order to elaborate his modal grammar of point of view, Simpson coined the notion of modal shading. In positive modal shading, the deontic and boulomaic modal systems are underlined, especially their higher degrees; this results in a text exposing the narrator's and their characters' duties, commitments and wishes, thereby signalling their mental and psychological features. A text with negative modal shading is typified by the epistemic and perception modal systems, generally their lower degrees; such a text indicates the narrator's and their characters' doubtfulness as to the veracity of the activities, events or other characters depicted. In neutral modal shading, modality is completely absent, and epistemically unmodalised categorical assertions, expressing the strongest possible degree of addresser commitment (Lyons 1977, 763, 808–809), are predominant; consequently, activities, events or characters are not psychologically appraised by the narrator or the other characters (Simpson 1993, 50–69; summarised in Neary 2014, 183–187).

Point of view has recently been connected with the idea of the disnarrated elaborated by Prince (1988; see also Lambrou 2019). The term indicates a break in the continuity and development of the plot; the reader may either be informed of this, or infer that this has taken place from a noteworthy gap in the narrative arrangement of events. Associated with the disnarrated is the concept of counterfactuality: this is constituted by a disnarrated (imagined or hypothetical) episode which is presented in place of the one actually belonging to the plot, and which meaningfully affects the assessment of certain aspects of the narrative. A counterfactual portrays a discourse world not presented as real, but which might have occurred and runs parallel to the real discourse world (Fauconnier and Turner 2002). As maintained by Riddle Harding (2007), counterfactuals are remarkable indicators of attitudinal position, as they urge the reader to take evaluative stances on the actualised and unactualised plot events.

Although only one of the texts scrutinised in this book (Text 2) features an essential, but stylistically relevant, instance of disnarration or counterfactuality (see Sect. 4.2), other ecological texts not explored here make more extensive use of this narratological device. A well-known example is the opening chapter of Rachel Carson's *Silent Spring* (1962) (see Sect. 2.1), which describes a town in the United States teeming with animal and vegetable life which, all of a sudden, dies out. The author contends that the town she depicts does not exist and the facts she relates did not actually happen, but adds that they are very likely to have equivalents in the country or anywhere else in the world. Similarly, several non-narrative and non-fictional environmental texts contain short narrative excerpts. Some of these excerpts concern the right options and courses of action not taken by us humans which would have safeguarded nature in the past; some other excerpts tell imagined stories and represent hypothetical scenarios about the doubtful future of our planet. The latter might be realised or unrealised, and might have desirable outcomes if we humans start preserving the natural world, or undesirable outcomes if we do not. Consequently, these disnarrated and counterfactual excerpts indicate their authors' attitudinal position to the environmental crisis, the reasons behind it and their possible consequences; as such, they make excellent linguistic data for further research in ecostylistics.[3]

2.2.3 Metaphor

Cognitive linguists Lakoff and Johnson, in their seminal book *Metaphors We Live By* (1980, second edition 2003), as well as Lakoff and Turner (1989) and the most recent cognitive researchers (such as Gibbs 2008; Kövecses 2020; Steen et al. 2010), argue that metaphors are not merely rhetorical devices, but mental and cultural phenomena "we live by". These authors write about "conceptual metaphor", because metaphor is conceived of as an attribute of concepts, not words. More precisely, in a conceptual metaphor a notion, domain or experience from a given semantic field (source domain or CONCEPT B, e.g. PLANTS) is used to conceptualise a notion, domain or experience from another semantic field (target domain or CONCEPT A, e.g. PEOPLE). The target is thereby understood in terms of the source in the resulting conceptual metaphor CONCEPT A IS CONCEPT B, here PEOPLE ARE PLANTS (see also Lakoff and Johnson's (2003) notion of mapping). The theoretical supposition is that the basic meanings of the source domain have little in common with those of the target domain: they "function as lexical points of entry to distinct conceptual source domains from which an indirect, metaphorical meaning is derived by means of a cross-domain mapping to the target domain" (Steen 2016, 315).

Why are stylistics and such related disciplines as semantics and literary criticism (among others) interested in metaphor, metaphor theory and figurative language? In Steen's (2014, 315) words, "The stylistic study of metaphor involves the idiosyncratic way metaphor is used in specific texts, by individual authors or, more broadly, sets of authors forming a school, generation or similar social groupings". The research purpose of stylistics is to provide a linguistic description of the text under examination, to interpret its meaning on the basis of that description and to evaluate the text and the message it communicates (Short 1996, 1–4). With regard to metaphor, stylistics therefore describes and interprets the cross-domain comparison between the target and the source, the reasons behind it, what distinctive semantic traits link the target to the source and what traits do not.

Conceptual metaphors are ubiquitous and pervasive features in several areas of human activity, including literature and the arts. Metaphor reconceptualises knowledge and experience by conveying the unfamiliar or abstract or complex (the target) through the familiar or concrete or simple (the source); in particular, literary writers can revitalise conceptual metaphors and make their use less automatic and conventionalised. Consequently, both novel and dead metaphors are regarded by Leech (2008) and van Peer (1986) as stylistic foregrounded deviations from the norm: they create defamiliarisation (Šklovsky 2012; see Sect. 2.2.1) and mainly, but not exclusively, characterise literary language.

Linguistic metaphorical expressions in literature are commonly imagined to be unique and unseen coinages of figurative equations. Nevertheless, Leech and van Peer maintain that the types of metaphor employed in literature may be the same as or similar to those employed in everyday language; Lakoff and Turner (1989) claim that underlying literary metaphors are conceptual metaphors which already exist at the culturally-shared conceptual level and may be very ordinary. Writers just subtly exploit and arrange them in masterly, ground-breaking and multi-layered images with artistic effect and deliberately-constructed ways of understanding the source and target of their metaphors. However, what is not generally the same or similar or ordinary is the degree of innovation of literary metaphors and the fact that, like the texts they occur in, they are intended to be re-read and re-visited: dwelling on and enjoying their various possible meanings are pivotal factors in literary criticism.

In Sperber and Wilson's (1995) relevance theory, utterances attempt to attain maximal efficiency and cognitive effect and, at the same time, to minimise processing efforts. Consequently, processing efforts are normally not minimal when, as previously mentioned, reading or re-reading literary texts and visiting or re-visiting innovative literary metaphors. It follows that one of the aims of the cognitive stylistic scrutiny of metaphor is to uncover the diverse processing depths of this figure when utilised in literature on the one hand and when utilised in everyday interaction on the other hand.

In cognitive stylistics, Goatly (1997) introduced the term "root analogies", which he defined as "metaphorical patterns which, on the lexicographical evidence, are important to conceptualization and cognition of English speakers" (Goatly and LLE Project 2002–2005). Root analogies are dissimilar to conceptual metaphors. While the latter are cognitive phenomena, the former are lexicographical phenomena, namely conventionalised and lexicalised metaphors. Moreover, while conceptual metaphors are usually composed of an abstract target and a concrete source (e.g. A RELATIONSHIP IS AN ANIMAL), the nature of target and source in root analogies can be at the same level of abstraction (e.g. HUMAN IS ANIMAL). The goal of cognitive stylistic investigation of literary and non-literary texts is to reveal that root analogies, even when realised in an inventive and refined way, underlie the patterns of imagery in these texts and contribute to highlighting their most meaningful motifs.

Rather than studying an isolated linguistic metaphorical expression or a single instance of a root analogy in a text, the cognitive approach, hence the cognitive stylistic investigation of metaphor, requires the analyst to distinguish extensive patterns or fields of metaphorical thought running through and structuring the entire text or discourse as connective forces. Werth (1994) applied his notion of megametaphor to the scrutiny of Dylan Thomas' *Under Milk Wood* and revealed that the play is unified by the frequent conceptual metaphors SLEEP IS DISABLEMENT and DISABLEMENT IS DEATH. Donald C. Freeman (1995) examined Shakespeare's *Macbeth* and the critical contributions about the drama; he demonstrated that the prevailing image schemata in both the language of the text and the critical language about the text are the CONTAINER schema and the PATH schema. Margaret H. Freeman (1995) disclosed that Emily Dickinson's poetic and conceptual world is constructed by the conceptual metaphor LIFE IS A VOYAGE IN SPACE, whereas a more frequent metaphor among her contemporaries was LIFE IS A JOURNEY THROUGH TIME.

Our language use activities (production, reception and interaction as a whole) are compelled by characteristic genre assumptions and our default knowledge about them. This is the background to one of the most recent developments in stylistic metaphor theory and analytical approach, viz. Steen's (2016) three-dimensional approach to metaphor. Three distinct dimensions are perceptible in all metaphor in all language use: (1) Expression, or the lexical, semantic and syntactic realisation of metaphor ("metaphor in language"); (2) Conceptualisation, or the representation of the target by means of the source and their cross-domain mapping ("metaphor in thought"); (3) Communication, or the deliberate or non-deliberate figurative function of metaphor ("metaphor in interaction"). This model, which is genre-driven, is based on "the three main components in any situation of discourse: the means of discourse, that is, language (expression), the individual discourse participants using their minds (conceptualization) and the combined participants doing their joint interaction (communication)" (Steen 2016, 310–311).

The notion of deliberate metaphor is key to this model. Moving from Shakespeare's Sonnet 18 "Shall I compare thee to a summer's day?", it is defined by Steen (2014, 316) as follows: "Deliberate metaphor is the intentional use of metaphor *as* metaphor, and its function is to offer an 'alien' perspective (a summer's day) on some target domain (the addressee in the poem as lover)". Metaphor is not very often utilised deliberately; when it is, the source domain and the target domain alike are explicitly underscored, and metaphor becomes a stylistic means foregrounded in the interaction-communication dimension. On the other hand, most metaphors are non-deliberate phenomena, a part of language-expression and thought-conceptualisation already existing in our linguistic and conceptual systems; hence, they are not remarkable from a stylistic viewpoint and in the interaction-communication dimension (Steen 2016, 318). In this book, in the five analytical Chapters 3–7, the metaphors found in the ecological texts under examination are compared with those catalogued in the three databases Master Metaphor List (Lakoff and Cognitive Linguistics Group 1991), Metalude (Goatly and LLE Project 2002–2005) and MetaNet Metaphor Wiki (David et al. 2016–2018); this is done with a view to proving whether these metaphors are more or less deliberate and creative.

Deliberate and non-deliberate metaphors are deployed differently in different registers, and the distinct genres and discourse events metaphor occurs in should be taken into consideration in stylistic and non-stylistic studies of this figure. Deliberate use of metaphor is most recurrent in the literary and poetic registers, where the potential of this figure is taken advantage of by conveying more than one simple conventional meaning; deliberate metaphor is also often present in advertising, news and political discourses, for the same reason and with the same function (Steen 2014: 317–319; 320; 324).

2.3 Ecostylistics

Thanks to its interdisciplinary inclination, presented in Sect. 2.2, stylistics can also effectively be wedded to disciplines in the humanities with ecological research agendas; as a consequence of this, ecostylistics is now an emerging field. The term "ecostylistics" was apparently first utilised by Andrew Goatly in his plenary speech "Edward Thomas, the landscape of nature, and ecostylistics", delivered at *PALA 2010: The Language of*

Landscapes, the thirtieth international conference of Poetics And Linguistics Association (PALA) (Genoa, Italy, July 2010); a written version of this plenary speech was published as Goatly (2017). In the introductory section of his speech and written contribution, this researcher firstly outlined the ideas in *The Wealth of Nations* (1776) by Adam Smith and contrasted them with the philosophy of Romantic Ecology analysed by Jonathan Bate (1991), which incorporates the concept of the pathetic fallacy developed by John Ruskin (1903). Afterwards, Goatly put forward the research question of whether it was possible to extend Romantic Ecology and ecological criticism closer to linguistics, i.e. to move towards an ecostylistics; with a view to giving a positive answer to this question, he deployed Systemic Functional Grammar to examine the noun phrases denoting nature in Edward Thomas' *Collected Poems*. In the concluding paragraph of his written contribution, this scholar notes: "the concept of this relatively new discipline [ecostylistics] can be extended further both in analysis of poetry, novels (see Chapter 10 of this volume by Elisabetta Zurru [2017]), and other media" (Goatly 2017, 121). According to Zurru (2017, 194), "Given Goatly's interest in both ecolinguistics (besides his 1996 article collected in the *Ecolinguistics Reader*, cf. e.g. Chapter 10 in his 2000 book *Critical Reading and Writing*) and stylistics (cf. Goatly 2008), it is not surprising that he should be the first to advocate an ecological turn in stylistic theory and practice".

Since Goatly's plenary speech, the concept of ecostylistics has indeed been extended in theory and practice to the study of other media and text-types, and interest in ecostylistic approaches has been growing internationally. New scholarly publications and conference presentations applying a stylistic methodology, rather than an ecocritical or ecolinguistic or discourse analytical paradigm, to landscape descriptions and to discourses about the environment are appearing more and more often, and ecostylistics is becoming a legitimate area in its own right. Authors working in the field are based at several institutions worldwide; the researchers most actively engaged in the area belong to or cooperate with LAND-SIG, a PALA Special Interest Group on the stylistics of landscape, place and environment coordinated by Ernestine Lahey

(University College Roosevelt), Elisabetta Zurru (University of Genoa) and myself (University of Cagliari).

Due to the wealth of the research interests of the environmental humanities on the one hand and of stylistics on the other hand, ecostylistics has so far addressed very different themes and wide-ranging issues. Zurru (2017, 195) observes that

> ecostylistics can be related to two main areas of investigation: the environmental and the ecological, respectively. [...] In other words, ecostylistics can focus either (a) on the link between the linguistic representation of physical environment and the style of a certain text, text-type, genre and/or author; (b) on the investigation and evaluation of (un)ecological linguistic patterns in texts, thus contributing to raising further awareness of global ecological concerns. The first aim is closer to the original subject of study of ecocriticism and appears to have been the main focus of ecostylistics so far (e.g. Lahey 2012; Virdis 2012). The second aim is closer to the early (and current) main scope of ecolinguistics and is deserving of careful consideration by ecostylisticians. (Zurru 2017, 195)

That is to say, stylistic scrutinies of depictions of natural or human-made scenery, usually in literary texts, have been prompted by ecocriticism; more political stylistic analyses criticising unecological discourses, usually in non-literary texts, have been prompted by ecolinguistics and may bring about academic activism in response to the environmental crisis. The research presented in this book is related to the latter, more committed, area of investigation inspired by the militant theoretical framework of ecolinguistics (see Sect. 2.1), and is carried out through the methodology of stylistics (see Sect. 2.2). Nevertheless, independently of the themes and issues being examined, the common thread underpinning all ecostylistic study is a careful reading of the texts and discourses under scrutiny, and an interpretation of their linguistic data and ecological aspects by means of the wide-ranging paradigms and approaches developed and supplied by contemporary literary and non-literary stylistics.

Recent book-length publications falling within the aims and scope of ecostylistics, or sharing similar research agendas, encompass the *Journal of World Languages* Special Issue on ecostylistics (Virdis, forthcoming), which features contributions applying the theoretical framework of stylistics to the examination of several literary and non-literary texts with ecological and environmental concerns. The edited collection *Language in Place: Stylistic Perspectives on Landscape, Place and Environment* (Virdis et al. 2021a) considers portrayals and conceptualisations of natural scenery and space in literary and non-literary texts by means of various stylistic methodologies. *Handbuch Sprache im urbanen Raum: Interdisziplinäre Perspektiven der Stadtforschung* (Busse and Warnke 2021), a handbook of language in urban spaces, treats the idea of urbanity as linguistically and communicatively shaped and presents urbanity as an interdisciplinary phenomenon. In the edited collection *Place-Making in the Declarative City* (Busse et al. 2020), the notion of the declarative city, referring to Searle's (1976) declarative speech act type, is examined from an interdisciplinary standpoint, ranging from linguistics, discourse analysis and literature to art history and urban planning. *Conceptualizing Biblical Cities: A Stylistic Study* (Vermeulen 2020) is a monograph contributing to the development of the stylistics of space in text and to the investigation of spatial construal in language by scrutinising the conceptual metaphors representing the city in the Bible. The monograph *Figura e ripetizione in* Dietro il paesaggio *di Andrea Zanzotto* (Cardilli 2020) is a stylistic study of Zanzotto's poetic language, principally of lexical and syntactic repetition and of the techniques utilised to depict the landscape and its geological and meteorological features. *The Stylistics of Landscapes, the Landscapes of Stylistics* (Douthwaite et al. 2017) is an edited collection of stylistic analyses of physical landscapes in literary texts, but also of stylistic examinations of metaphorical landscapes in literature, viz. social, mental and historical portrayals of places, people and society. Although not adhering to a stylistic methodology but mainly to ecolinguistics and positive discourse analysis, the monograph *Discourses of Cycling, Road Users and Sustainability: An Ecolinguistic Investigation* (Caimotto 2020) studies ecological matters with the same aim as this book: to identify and spread beneficial discourses, discursive strategies and everyday practices.

A number of article-length publications and shorter contributions, not comprised in the volumes cited above, also investigate natural and anthropogenic scenery from the disciplinary viewpoint of stylistics. Chrzanowska-Kluczewska (2020) outlines the predominant stylistic tools employed to describe the physical environments in Mary Shelley's *Frankenstein*, and proves that the natural settings in the novel are associated with the eponymous character's emotional temperament. This researcher (Chrzanowska-Kluczewska 2012) also explores the landscapes in a selection of poems by Czesław Miłosz through the apparatus of Text World Theory and points out the existence of regular text worlds behind a body of poetry. Goatly (2008) discusses several poems from Alfred Edward Housman's *A Shropshire Lad* by investigating, from a Systemic Functional Grammatical perspective, the semantic content of the marked Themes figuring in them, and the connection of these linguistic marked Themes with the literary themes of place, time and exile; this author elaborates this research and the adoption of Systemic Functional Grammar to scrutinise phrases referring to place, direction and the natural world in his book chapters Goatly (2021) and Goatly (2017). Leech and Short (2007, 66–72) analyse a text depicting a waterscape (the opening passage of Joseph Conrad's *The Secret Sharer*) by pinpointing its style markers (lexical and grammatical categories, figures of speech, context and cohesion); they show that these markers communicate the narrator's exploring his alien surroundings and trying to make sense of them. Werth (1999, 7–10, 52–60) examines extracts from Edward Morgan Foster and Ernest Hemingway portraying fictional spaces; he demonstrates that graphical representations of the spaces described in the extracts must be similar to the mental representations underlying our processing of the extracts; he also demonstrates that fictional locations and text worlds begin building up by means of just a few basic deictic and referential world-defining elements.

The work of several other authors furnishes further instances of the different themes and issues tackled by ecostylistics. Virdis studies metaphor and framing in two ecologically-aware lectures by John Ruskin (Virdis 2022; see also Virdis and Cadeddu, forthcoming); the Systemic Functional Grammatical practices to represent cats on the website of a charity (Virdis 2021) and birds in Richard Bach's *Jonathan Livingston Seagull* (Virdis 2015); the stylistic mechanisms to depict natural scenery as sexualised in a Victorian pornographic novel (Virdis 2019; see also

Virdis and Milia 2021); the linguistic strategies to portray the physical environment as cultural heritage in a tourism website (Virdis 2016); and the grammatical techniques to describe the Indian countryside in a colonial text (Virdis 2012). Vermeulen investigates the stylistic and conceptual construction of cities as homes in biblical Psalm 137 and in three Antwerp city poems, questioning the spatial categories of "city" and "home" (Vermeulen 2017a); the depiction of Nineveh in Jonah 3, drawing on a cognitive stylistic analysis of space builders and the spatial conceptualisation of the city (Vermeulen 2017b); and the construction of the enemy city in the Hebrew Bible and the textual space-producing process through Lefebvre's (1974) and Soja's (1996) spatial concept of Thirdspace (Vermeulen 2016). Alarcón-Hermosilla (2017) scrutinises the manipulation of language to foreground the role of nature, light and darkness and their association with religion and death in John McGahern's *The Dark* by adhering to Talmy's (2000) principle of Fictive Motion and to conceptual metaphor theory. Pelclová (2017) categorises the urban signage, or written verbal expressions, in a corpus of texts collected in Brno, Czech Republic, into Jakobson's (1960) six functions of language, thereby showing how diverse urban spaces and identities are discursively constructed. Zurru explores the official tourism website of the Maltese archipelago and reveals that it constantly portrays the landscape of the islands and their historical and cultural heritage as authentically Mediterranean (Zurru 2016); this scholar also analyses the cohesive and transitivity patterns utilised in Amitav Ghosh's *The Hungry Tide* to consider ecological issues (Zurru 2012). Lahey examines Al Purdy's *North of Summer* and the Canadian Arctic landscapes it represents through conceptual integration theory (Lahey 2012); this researcher also applies Text World Theory to study twentieth-century Canadian poetic discourse and the description of Canadian natural scenery appearing in this discourse, in particular how this poetry functions stylistically by means of megametaphors, or key metaphorical patterns (Lahey 2007); the construction of text-world locations with real-world counterparts (Lahey 2006); and the interplay between external (viz. physical) environments and internal (viz. mental) environments (Lahey 2003).

Finally, the ecostylistic paradigm has been recently adopted in Translation Studies to investigate the translation of space and landscape and the patterns revealing the translator's approach to nature and natural scenery. Fois extensively scrutinises the English version of Grazia Deledda's *La madre*, more specifically anthropomorphic metaphor as a tool to depict the physical environment as a moral agent worthy of respect, and its translation as a means to express the relationship between the environment and cultural identity (Fois 2022); the impact of the translator's cultural framework on the translation of landscape-related metaphor (Fois 2020a); how the translator's interpretation of natural scenery in the source text, especially evident in the conversion of metaphor into simile, can reshape the target readers' understanding of it (Fois 2020b); and the translational choices linked to space, nature and the physical environment as features affecting the reception of foreign authors (Fois 2020c).

Thanks to this broad range of themes and issues, the audience for ecostylistics is wide-ranging. It is mostly composed of academics researching, teaching or interested in the disciplines of ecolinguistics, stylistics and ecocriticism. The themes and issues considered by ecostylistics also appeal to scholars of such neighbouring subject and research areas as discourse analysis, environmental communication and environmental humanities. Furthermore, since ecostylistics examines ecological discourses in distinct author-types and text-types, it also attracts researchers focusing their attention on how those author-types and text-types deal with ecological matters. Among the target audience of ecostylistics, particularly of the non-analytical (i.e. less specialised and technical) sections of its publications, is also the general public with an interest in ecological topics and with environmental concerns—for instance, members of and people working for ecological non-governmental and intergovernmental organisations; people interested in environmental activities and campaigns; users of websites and social media discussing ecological issues; school teachers educating their students in environmentalism; journalists popularising ecological subjects.

As shown above, ecostylistics is particularly concerned with the stylistic analysis of natural and anthropogenic space and scenery in various text-types; as a result, it contributes to the scholarly debate on landscape, the environment and its human and animal dwellers currently taking place in the humanities and social sciences. This debate has followed and evolved from the cultural turn in geography in the 1980s (which holds landscape, place and environment to be changing, multi-layered and culturally-influenced notions), from the spatial turn in the humanities in the same years (which tends towards spatial thinking rather than temporal thinking, and synchronic methods rather than diachronic methods) and from the academic acknowledgement of the ecological turn in the social sciences and the humanities a decade later (whose priorities are environmental risk, the balance between human life and more-than-human life, and the environmental crisis). These "turns" inspired the birth of the environmental humanities as a whole and several disciplines with ecological concerns, ranging from ecofeminism (Adams and Gruen 2014) and ecopsychology (Roszak et al. 1995) to ecopoetics (Knickerbocker 2012) and ecotranslation (Cronin 2017) (in addition, see Barnes and Duncan 1992; Bekoff 2007; Blunt and Rose 1994; Braddock and Irmscher 2009; Cosgrove 1985; Garrard 2014b; Herman 2016; Mitchell 2002; Rose 1993, 1995; Westphal 2011. For the spatial turn and its relevance for stylistics, see Virdis et al. 2021b; for a recent study of the relations among landscape, language, culture and geography, see Bromhead 2018).

One of the disciplines leading this debate is ecocriticism: it supplies an ecological viewpoint on literary, media and popular culture studies, and often underlines aesthetic, ethical and activist preoccupations in the texts it explores. It is generally agreed among scholars that an ecological turn within academia occurred in the 1970s, but contemporary ecocriticism can be traced back to Leo Marx's (1964) book *The Machine in the Garden: Technology and the Pastoral Ideal in America*. The field was given its name at the end of the 1970s, when William Rueckert (1978) employed the term in the title of one of his articles; it was academically acknowledged as a flourishing area in the mid-1990s, when Cheryll Glotfelty and Harold Fromm (1996) published *The Ecocriticism Reader: Landmarks in Literary Ecology* (for the distinction between first-wave and second-wave ecocriticism and the social and ecological concerns of the two waves, see Garrard 2014a and Mukherjee 2007).

Another discipline actively involved in the ecological debate is environmental communication which, unlike ecocriticism, is inclined to examine concrete matters in environmental journalism and the sociology of ecological movements. Among the first principal publications in environmental communication are M. Jimmie Killingsworth and Jacqueline S. Palmer's (1992) *Ecospeak: Rhetoric and Environmental Politics in America* and James G. Cantrill and Christine L. Oravec's (1996) *The Symbolic Earth: Discourse and Our Creation of the Environment*. The analytical aim of the former book is to determine the rhetorical structures commonly utilised in written discourse about environmental politics, whereas the latter work centres on public and political discourses like environmental advocacy, green consumerism and environmental reporting (for the historical development of and the existing trends in both ecocriticism and environmental communication, and for the emergence of the meta-discipline of environmental humanities, see Slovic et al. 2019).

Given the scholarly role taken by ecocriticism and environmental communication, and the wider academic attention aroused in the humanities and social sciences by ecological topics, a number of book series dealing with these topics have been published; a few volumes share certain research interests with ecostylistics. For instance, "Palgrave Studies in Media and Environmental Communication", edited by Anders Hansen and Steve Depoe, centres on the broad and cross-disciplinary field of environmental communication. In other words, its authors investigate the principal communicative strategies and mechanisms deployed at the macro-level of the media to cover local and global ecological matters, dangers and calamities. Despite being founded on an interdisciplinary theoretical framework and methodology, ecostylistics furnishes more focused research on the ecostylistic practices and processes used at the micro-level of the individual texts under investigation. The book series "Literatures, Cultures, and the Environment", edited by Ursula K. Heise, consists of volumes in the subject area of environmental humanities, especially literary or rhetorical research. The volumes scrutinise how distinct literatures and cultures at distinct points in time shaped concepts of nature and ecological preoccupations. Although one of the goals of a number of publications in ecostylistics is to define such keywords as nature, ecology and space in various texts and discourses, ecostylistic publications do not scrutinise literary texts only or adopt the paradigm of cultural or rhetorical studies: they explore newer and older literary and non-literary texts by means of ecostylistics. "Environmental Cultures", a book series edited by Richard Kerridge and Greg Garrard, features literary ecocriticism treating the different cultural reactions to the environmental crisis. The work in the series covers a number of media, disciplinary fields and geographical areas, and goes beyond more mainstream forms of cultural study. Ecostylistics also explores our various human associations with the earth, crosses a number of disciplines (for instance, ecolinguistics and stylistics), text-types and languages, and is ground-breaking in its providing stylistics with ecological preoccupations; in contrast to this book series, however, ecostylistics does not exclusively analyse literary texts, and does not adhere to an ecocritical approach, but to a linguistic approach even when examining literature.

Among the diverse analytical positions listed above, this book adopts a critical stance and is in line with the scholarly and social militancy pursued by a part of ecostylistics and by ecolinguistics. It investigates the ecological discourse conveyed by non-literary evidence by supplying a careful and detailed ecostylistic analysis of specific texts and a comprehensive study of their primary distinctive linguistic traits in accordance with the approach of ecostylistics. Its aim is to have a positive practical impact on the ecostylistic techniques deployed to communicate ecological matters: it recognises alterative beneficial models and promotes a broader use and dissemination of life-sustaining ecosophies and ecological agendas. To be more precise, compared with the relevant publications and book series mentioned above, this book linguistically analyses only contemporary texts belonging to non-literary text-types and treats exclusively topical ecological subjects in up-to-date ecologically-aware linguistic data. It extends the scope of the investigation from the scrutiny of landscape in particular to the scrutiny of nature, the environment and sustainability in general; no metaphorical or figurative scenery is taken into consideration. It is ecologically committed and, in the following analytical Chapters 3–7, studies texts and discourses declaring themselves to be environment-oriented.

Notes

1. Harré et al. (1999), who "believe that ecological and environmental studies need to take a linguistic turn" (Harré et al. 1999, 1), elaborated a model of language planning bearing ecological and social issues in mind, and organised into four criteria to appraise the adequacy of a language: "*Referential adequacy* means 'the capacity of the language to meet the needs of its users as an instrument of referential meaning' (Haugen 1966, 62). *Systematic adequacy* means 'being structured so as to approach maximum rule economy and efficiency, and having a clear and uniform semantic structure with a terminology that is unambiguously translatable' (Dahlstedt 1979, 27). *Social adequacy* means that language should be acceptable to a maximum number of speakers in the target community, promote social unity and intercommunication and cater to present as well as anticipated future social needs. *Environmental adequacy* means that a language should enable its users to talk about environmental matters in an informed manner and promote the well-being of its speakers and nonhuman nature" (Harré et al. 1999, 22).
2. Chawla (1991) also contended that, unlike Amerindian languages, English and the worldview and habits associated with it are ill-suited to ecological and holistic principles, given that the "linguistic and philosophical roots of our environmental crisis", as the title of her article indicates, are founded on three characteristics and strategies of the English language emphasising things and not processes: (1) individualising mass nouns in such phrases as "a mug of beer", "a cup of coffee", "a glass of water", thus fragmenting entities and the wholeness of life; (2) counting intangible entities like quality of life and sense of wellbeing by means of indicators, thereby fragmenting our human needs and objectifying and measuring our human experience; (3) the three-tense time scheme, which objectifies time, makes it linear and splits it up into equal units, while obliterating our subjective experience of it and sustaining the ideology of progress.
3. A number of devices expressing point of view mostly occur in narrative and fictional text-types, and are not present in the non-narrative and non-fictional texts investigated in this book; therefore, they are not discussed in depth here. Among them are speech and thought presentation and mindstyle. Point of view interconnects with and is influenced by the narrative strategy of speech and thought presentation, which consists of a narrator presenting the discourse of a character. The common ground shared by the stylistic practices of point of view and discourse presentation

is the speech and thought presentation category of free indirect discourse (FID), which comprises the variants of speech (FIS) and thought (FIT). This is a primary category, for it produces the effect that the narrator and the character are speaking at the same time, and that their various voices are merged (Simpson 2010, 296–297). A notion introduced by Fowler (1977, 103), mindstyle denotes the specific linguistic construction of a narrator's or character's mental individuality. Grammatical and lexical selections signal a narrator's idiosyncratic, and frequently somewhat limited, cognitive standpoint; as a result, the plot unfolds from the restricted viewpoint of this peculiar narrator, and the reader is required to look outside the range of the narrator's partial understanding and mindset (Simpson 2010, 298–299). For the sociopragmatic and cognitive features suggesting point of view present *around* the text, not *in* the text, see Simpson (2010, 301–309); for point of view in dramatic texts, see McIntyre (2006).

References

Adams, C.J., and L. Gruen, eds. 2014. *Ecofeminism: Feminist Intersections with Other Animals and the Earth*. London: Bloomsbury.

Alarcón-Hermosilla, S. 2017. Fictive motion in John McGahern. Light and Nature in *The Dark*. *Fictions: Studi sulla narratività* 16: 91–100.

Baker, S. 2006. *Sustainable Development*. London and New York: Routledge.

Barnes, T.J., and J.S. Duncan, eds. 1992. *Writing Worlds: Discourse, Text and Metaphor in the Representation of Landscape*. London and New York: Routledge.

Bate, J. 1991. *Romantic Ecology: Wordsworth and the Environmental Tradition*. London and New York: Routledge.

Bekoff, M., ed. 2007. *Encyclopedia of Human-Animal Relationships: A Global Exploration of Our Connections with Animals*. Santa Barbara, CA: Greenwood.

Blunt, A., and G. Rose, eds. 1994. *Writing Women and Space: Colonial and Postcolonial Geographies*. New York and London: The Guilford Press.

Braddock, A.C., and C. Irmscher, eds. 2009. *A Keener Perception: Ecocritical Studies in American Art History*. Tuscaloosa, AL: University of Alabama Press.

Bromhead, H. 2018. *Landscape and Culture: Cross-Linguistic Perspectives*. Amsterdam and Philadelphia: John Benjamins.

Burke, M., ed. 2014. *The Routledge Handbook of Stylistics*. London and New York: Routledge.

Busse, B., and I.H. Warnke, eds. 2021. *Handbuch Sprache im urbanen Raum: Interdisziplinäre Perspektiven der Stadtforschung*. Berlin and Boston, MA: de Gruyter.

Busse, B., I.H. Warnke, and J. Smith, eds. 2020. *Place-Making in the Declarative City*. Berlin and Boston, MA: de Gruyter.

Caimotto, M.C. 2020. *Discourses of Cycling, Road Users and Sustainability: An Ecolinguistic Investigation*. Basingstoke: Palgrave Macmillan.

Cantrill, J.G., and C.L. Oravec. 1996. *The Symbolic Earth: Discourse and Our Creation of the Environment*. Lexington, KY: University Press of Kentucky.

Capra, F. 1996. *The Web of Life: A New Scientific Understanding of Living Systems*. New York: Anchor Books.

Cardilli, L. 2020. *Figura e ripetizione in* Dietro il paesaggio *di Andrea Zanzotto*. Milan and Udine: Mimesis Edizioni.

Carson, R. 1962. *Silent Spring*. Boston, MA: Houghton Mifflin Company.

Carter, R. 2010. Methodologies for stylistic analysis. In *Language and Style: In Honour of Mick Short*, eds. D. McIntyre and B. Busse, 55–68. Basingstoke: Palgrave Macmillan.

Chawla, S. 1991. Linguistic and philosophical roots of our environmental crisis. *Environmental Ethics* 13 (3): 253–273.

Chrzanowska-Kluczewska, E. 2012. "Remembrance of things past"—The lost landscapes of Marcel Proust and Czesław Miłosz. In *Poetry and Its Language: Papers in Honour of Teresa Bela*, ed. M. Gibińska and W. Witalisz, 217–233. Frankfurt am Main: Peter Lang.

Chrzanowska-Kluczewska, E. 2020. The aesthetics of *Frankenstein*: Nightmarish landscapes in European Romantic and symbolic fantasy literature and fantastic painting. In *Frankenstein 1818–2018: Parabel der Moderne / Parable of the Modern Age*, ed. D. Petzold and K. Seibel, 73–90. Berlin: Peter Lang.

Cook, G., and A. Sealey. 2018. The discursive representation of animals. In *The Routledge Handbook of Ecolinguistics*, ed. A.F. Fill and H. Penz, 311–324. London and New York: Routledge.

Cosgrove, D. 1985. Perspective and the evolution of the landscape idea. *Transactions of the Institute of British Geographers, New Series* 10 (1): 45–62.

Coupe, L., ed. 2000. *The Green Studies Reader: From Romanticism to Ecocriticism*. London and New York: Routledge.

Crompton, T. 2010. *Common Cause: The Case for Working with Our Cultural Values*. https://assets.wwf.org.uk/downloads/common_cause_report.pdf. Last accessed April 2022.
Cronin, M. 2017. *Eco-Translation: Translation and Ecology in the Age of the Anthropocene*. London and New York: Routledge.
Dahlstedt, K.H. 1979. Societal ideology and language cultivation. *Linguistics* 183: 17–50.
David, O., G. Lakoff, et al. 2016–2018. *MetaNet*. Berkeley: University of California. https://metaphor.icsi.berkeley.edu/pub/en. Last accessed April 2022.
Douthwaite, J., D.F. Virdis, and E. Zurru, eds. 2017. *The Stylistics of Landscapes, the Landscapes of Stylistics*. Amsterdam and Philadelphia: John Benjamins.
Drengson, A. 1995. Shifting paradigms: From technocrat to planetary person. In *The Deep Ecology Movement: An Introductory Anthology*, ed. A. Drengson and Y. Inoue, 74–100. Berkeley, CA: North Atlantic Books.
Drengson, A., and Y. Inoue. 1995. *The Deep Ecology Movement: An Introductory Anthology*. Berkeley, CA: North Atlantic Books.
Ehrlich, P.R. 1968. *The Population Bomb*. San Francisco and New York: Sierra Club and Ballantine Books.
Emmott, C., and M. Alexander. 2014. Foregrounding, burying and plot construction. In *The Cambridge Handbook of Stylistics*, ed. P. Stockwell and S. Whiteley, 329–343. Cambridge: Cambridge University Press.
Emmott, C., and M. Alexander. 2016. Defamiliarization and foregrounding: Representing experiences of change of state and perception in neurological illness autobiographies. In *The Bloomsbury Companion to Stylistics*, ed. V. Sotirova, 289–307. London: Bloomsbury.
Fairclough, N. 2010 [1995]. *Critical Discourse Analysis: The Critical Study of Language*, 2nd ed. Harlow: Longman.
Fauconnier, G., and M. Turner. 2002. *The Way We Think: Conceptual Blending and the Mind's Hidden Complexities*. New York: Basic Books.
Fill, A.F. 2018. Introduction. In *The Routledge Handbook of Ecolinguistics*, ed. A.F. Fill and H. Penz, 1–7. London and New York: Routledge.
Fill, A.F., and H. Penz, eds. 2018a. *The Routledge Handbook of Ecolinguistics*. London and New York: Routledge.
Fill, A.F., and H. Penz. 2018b. Ecolinguistics in the 21st century: New orientations and future directions. In *The Routledge Handbook of Ecolinguistics*, ed. A.F. Fill and H. Penz, 437–443. London and New York: Routledge.

Fill, A.F., and P. Mühlhäusler, eds. 2001. *The Ecolinguistics Reader: Language, Ecology, and Environment*. London and New York: Continuum.

Fois, E. 2020a. The English translation of Grazia Deledda's *La madre* and the relevance of culture in translating landscape metaphor. *Cadernos de Tradução* 40 (2): 112–130.

Fois, E. 2020b. The rendition of metaphors and the translator's influence in the English translation of Grazia Deledda's *La madre*. *Palimpsest: International Journal for Linguistic, Literary and Cultural Research* 5: 53–66.

Fois, E. 2020c. Translating foreign space and landscape: *The Mother* by Grazia Deledda. *InVerbis: Lingue Letterature Culture* 10 (2): 137–160.

Fois, E. 2022. Translating anthropomorphic metaphors: An ecostylistic contrastive analysis of Grazia Deledda's *The Mother*. *International Journal of English Linguistics* 12 (1): 1–11.

Fowler, R. 1977. *Linguistics and the Novel*. London: Methuen.

Fowler, R. 1986. *Linguistic Criticism*. Oxford: Oxford University Press.

Fowler, R. 1996 [1986]. *Linguistic Criticism*, 2nd ed. Oxford: Oxford University Press.

Freeman, D.C. 1995. Catch[ing] the nearest way: Macbeth and cognitive metaphor. *Journal of Pragmatics* 24: 689–708.

Freeman, M.H. 1995. Metaphor making meaning: Dickinson's conceptual universe. *Journal of Pragmatics* 24: 643–666.

Gare, A. 1996. *Nihilism Inc.: Environmental Destruction and the Metaphysics of Sustainability*. Como, NSW: Eco-Logical Press.

Garrard, G. 2014a. Introduction. In *The Oxford Handbook of Ecocriticism*, ed. G. Garrard, 1–24. Oxford: Oxford University Press.

Garrard, G., ed. 2014b. *The Oxford Handbook of Ecocriticism*. Oxford: Oxford University Press.

Gavins, J. 2014. Defamiliarisation. In *The Cambridge Handbook of Stylistics*, ed. P. Stockwell and S. Whiteley, 196–211. Cambridge: Cambridge University Press.

Gibbs, R.W., Jr., ed. 2008. *The Cambridge Handbook of Metaphor and Thought*. Cambridge: Cambridge University Press.

Glotfelty, C., and H. Fromm, eds. 1996. *The Ecocriticism Reader: Landmarks in Literary Ecology*. Athens, GA: The University of Georgia Press.

Goatly, A. 1996. Green grammar and grammatical metaphor, or language and myth of power, or metaphors we die by. *Journal of Pragmatics* 25: 537–560.

Goatly, A. 1997. *The Language of Metaphors*. London: Routledge.

Goatly, A. 2000. *Critical Reading and Writing: An Introductory Coursebook*. London and New York: Routledge.

Goatly, A. 2008. Marked Theme and its interpretation in A.E. Housman's *A Shropshire Lad*. In *Explorations in Stylistics*, 59–96. Sheffield: Equinox.

Goatly, A. 2017. The poems of Edward Thomas: A case study in ecostylistics. In *The Stylistics of Landscapes, the Landscapes of Stylistics*, ed. J. Douthwaite, D.F. Virdis, and E. Zurru, 95–122. Amsterdam and Philadelphia: John Benjamins.

Goatly, A. 2021. Thematic adverbial adjuncts of place and direction and their relationship to conceptual metaphor in A.E. Housman's *A Shropshire Lad*. In *Language in Place: Stylistic Perspectives on Landscape, Place and Environment*, ed. D.F. Virdis, E. Zurru, and E. Lahey, 17–44. Amsterdam and Philadelphia: John Benjamins.

Goatly, A., and LLE Project. 2002–2005. *Metalude*. Hong Kong: Lingnan University. http://www.ln.edu.hk/lle/cwd/project01/web/home.html. Last accessed April 2022.

Gobard, H. 1976. *L'Aliénation Linguistique: Analyse Tétraglossique*. Paris: Flammarion.

Gregoriou, C. 2014. The linguistic levels of foregrounding in stylistics. In *The Routledge Handbook of Stylistics*, ed. M. Burke, 87–100. London and New York: Routledge.

Haeckel, E. 1866. *Generelle Morphologie der Organismen. Allgemeine Grundzüge der Organischen Formen-Wissenschaft, Mechanisch Begründet durch die von Charles Darwin Reformirte Descendenztheorie*. Berlin: G. Reimer.

Halliday, M.A.K. 1990. New ways of meaning: The challenge to applied linguistics. *Journal of Applied Linguistics* 6: 7–36.

Handl, G. 2012. *Declaration of the United Nations Conference on the Human Environment (Stockholm Declaration), 1972 and the Rio Declaration on Environment and Development, 1992*. United Nations Audiovisual Library of International Law. https://legal.un.org/avl/pdf/ha/dunche/dunche_e.pdf. Last accessed April 2022.

Harré, R., J. Brockmeier, and P. Mühlhäuser. 1999. *Greenspeak: A Study of Environmental Discourse*. London: Sage.

Harrison, K.D. 2007. *When Languages Die: The Extinction of the World's Languages and the Erosion of Human Knowledge*. Oxford: Oxford University Press.

Haugen, E. 1966. Linguistics and language planning. In *Sociolinguistics*, ed. W. Bright, 50–71. The Hague: Mouton.

Haugen, E. 1972. The ecology of language. In *The Ecology of Language: Essays by Einar Haugen*, ed. A.S. Dil, 325–339. Stanford, CA: Stanford University Press.

Herman, D., ed. 2016. *Creatural Fictions: Human-Animal Relationships in Twentieth- and Twenty-First-Century Literature*. Basingstoke: Palgrave Macmillan.

Heuberger, R. 2018. Overcoming anthropocentrism with anthropomorphic and physiocentric uses of language? In *The Routledge Handbook of Ecolinguistics*, ed. A.F. Fill and H. Penz, 342–354. London and New York: Routledge.

Howlett, M., and R. Raglon. 1992. Constructing the environmental spectacle: Green advertisements and the greening of the corporate image, 1910–1990. *Environmental History Review* 16 (4): 53–68.

Hu, (H.)G. 2020. *Eco-Translatology: Towards an Eco-Paradigm of Translation Studies*. Singapore: Springer.

Hummelsund Voie, C. 2019. Nature writing in the Anthropocene. In *Routledge Handbook of Ecocriticism and Environmental Communication*, ed. S. Slovic, S. Rangarajan, and V. Sarveswaran, 199–210. London and New York: Routledge.

Ihlen, Ø. 2009. Business and climate change: The climate response of the world's 30 largest corporations. *Environmental Communication: A Journal of Nature and Culture* 3 (2): 244–262.

Jakobson, R. 1960. Closing statement: Linguistics and poetics. In *Style in Language*, ed. T. Sebeok, 350–377. Cambridge, MA: MIT Press.

Jeffries, L., and D. McIntyre. 2010. *Stylistics*. Cambridge: Cambridge University Press.

Killingsworth, M.J., and J.S. Palmer. 1992. *Ecospeak: Rhetoric and Environmental Politics in America*. Carbondale, IL: Southern Illinois University Press.

Killingsworth, M.J., and J.S. Palmer. 2019. Foreword. In *Routledge Handbook of Ecocriticism and Environmental Communication*, eds. S. Slovic, S. Rangarajan, and V. Sarveswaran, xviii–xxi. London and New York: Routledge.

Knickerbocker, S. 2012. *Ecopoetics: The Language of Nature, the Nature of Language*. Amherst, MA: University of Massachusetts Press.

Kövecses, Z. 2020. *Extended Conceptual Metaphor Theory*. Cambridge: Cambridge University Press.

Lahey, E. 2003. Seeing the forest for the trees in Al Purdy's "Trees at the Arctic Circle". *BELL: Belgian Journal of English Language and Literatures* 1: 73–83.

Lahey, E. 2006. (Re)thinking world-building: Locating the text-worlds of Canadian lyric poetry. *Journal of Literary Semantics* 35 (2): 145–164.

Lahey, E. 2007. Megametaphorical mappings and the landscapes of Canadian poetry. In *Contemporary Stylistics*, ed. M. Lambrou and P. Stockwell, 156–167. London: Continuum.
Lahey, E. 2012. Blended discourses: Reading the multimodal North in Al Purdy's *North of Summer*. In *Contemporary Canadian Literature: European Approaches*, ed. D. Zetu, 68–99. Iasi: Cuza University Press.
Lakoff, G. 2010. Why it matters how we frame the environment. *Environmental Communication* 4 (1): 70–81.
Lakoff, G., and Cognitive Linguistics Group. 1991. *Master Metaphor List*. Berkeley: University of California. http://araw.mede.uic.edu/~alansz/metaphor/METAPHORLIST.pdf. Last accessed April 2022.
Lakoff, G., and M. Johnson. 2003 [1980]. *Metaphors We Live By: With a New Afterword*, 2nd ed. Chicago: University of Chicago Press.
Lakoff, G., and M. Turner. 1989. *More than Cool Reason: A Field Guide to Poetic Metaphor*. Chicago: University of Chicago Press.
Lambrou, M. 2019. *Disnarration and the Unmentioned in Fact and Fiction*. Basingstoke: Palgrave Macmillan.
Leech, G. 2008. *Language in Literature: Style and Foregrounding*. London: Pearson.
Leech, G., and M. Short. 2007 [1981]. *Style in Fiction: A Linguistic Introduction to English Fictional Prose*, 2nd ed. Harlow: Longman.
Lefebvre, H. 1974. *La Production de l'Espace*. Paris: Editions Anthropos.
Lyons, J. 1977. *Semantics*, 2 vols. Cambridge: Cambridge University Press.
Macfarlane, R. 2013. New words on the wild. *Nature* 498: 166–167.
Marcellesi, J.-B. 1975. Basque, breton, catalan, corse, flamand, germanique d'Alsace, occitan: L'enseignement des 'langues régionales.' *Langue Française* 25: 3–11.
Marx, L. 1964. *The Machine in the Garden: Technology and the Pastoral Ideal in America*. Oxford: Oxford University Press.
Masami, Y. 2019. Designing the communication of traditional ecological knowledge: A Noto case study. In *Routledge Handbook of Ecocriticism and Environmental Communication*, ed. S. Slovic, S. Rangarajan, and V. Sarveswaran, 279–290. London and New York: Routledge.
Mautner, G. 2010. *Language and the Market Society: Critical Reflections on Discourse and Dominance*. London: Routledge.
McIntyre, D., 2006. *Point of View in Plays*. Amsterdam and Philadelphia: John Benjamins.
Mitchell, W.J.T. 2002. Imperial landscape. In *Landscape and Power*, ed. W.J.T. Mitchell, 5–34. Chicago, IL and London: University of Chicago Press.

Mühlhäusler, P. 2000. Bleached language on unbleached paper: The language of ecotourism. In *ECOnstructing Language, Nature and Society: The Ecolinguistic Project Revisited: Essays in Honour of Alwin Fill*, ed. B. Ketteman and H. Penz, 241–253. Tübingen: Stauffenburg Verlag.

Mühlhäusler, P., and A. Peace. 2006. Environmental discourses. *Annual Review of Anthropology* 35: 457–479.

Mukařovský, J. 1964. Standard language and poetic language. In *A Prague School Reader on Esthetics, Literary Structure and Style*, ed. P.L. Garvin, 17–30. Washington, D.C.: Georgetown University Press.

Mukařovský, J. 1977. On poetic language. In *The Word and Verbal Art: Selected Essays*, ed. J. Burbank and P. Steiner, 1–64. New Haven, CT and London: Yale University Press.

Mukherjee, U.P. 2007. Surfing the second wave: Amitav Ghosh's tide country. *New Formations* 59: 144–157.

Myerson, G., and Y. Rydin. 1996. *The Language of Environment: A New Rhetoric*. London and New York: Routledge.

Naess, A. 1995a. The shallow and the deep, long-range ecology movement: A summary. In *The Deep Ecology Movement: An Introductory Anthology*, ed. A. Drengson and Y. Inoue, 3–9. Berkeley, CA: North Atlantic Books.

Naess, A. 1995b. The systematization of the logically ultimate norms and hypotheses of Ecosophy T. In *The Deep Ecology Movement: An Introductory Anthology*, ed. A. Drengson and Y. Inoue, 31–48. Berkeley, CA: North Atlantic Books.

Neary, C. 2014. Stylistics, point of view and modality. In *The Routledge Handbook of Stylistics*, ed. M. Burke, 175–190. London and New York: Routledge.

Olson, C. 2000. *Zen and the Art of Postmodern Philosophy*. Albany, NY: State University of New York Press.

Orr, D. 1992. *Ecological Literacy: Education and the Transition to a Postmodern World*. New York: SUNY Press.

Oxford English Dictionary (OED) Online. 2022. Oxford: Oxford University Press. https://www.oed.com/. Last accessed April 2022.

Pelclová, J. 2017. Signifying Brno: Creating urban space, shaping the city. *Brno Studies in English* 43 (1): 69–87.

Prince, G. 1988. The disnarrated. *Style* 22: 1–8.

Quirk, R., S. Greenbaum, G. Leech, and J. Svartvik. 1985. *A Comprehensive Grammar of the English Language*. Harlow: Longman.

Riddle Harding, J. 2007. Evaluative stance and counterfactuals in language and literature. *Language and Literature* 16 (3): 263–280.

Rose, G. 1993. *Feminism and Geography: The Limits of Geographical Knowledge*. Cambridge: Polity Press.

Rose, G. 1995. Distance, surface, elsewhere: A feminist critique of the space of phallocentric self/knowledge. *Environment and Planning d: Society and Space* 13: 761–781.

Roszak, T., M.E. Gomes, and A.D. Kanner, eds. 1995. *Ecopsychology: Restoring the Earth, Healing the Mind*. San Francisco: Sierra Club.

Rothenberg, D. 1995. A platform of deep ecology. In *The Deep Ecology Movement: An Introductory Anthology*, ed. A. Drengson and Y. Inoue, 155–166. Berkeley, CA: North Atlantic Books.

Rueckert, W. 1978. Literature and ecology: An experiment in ecocriticism. *Iowa Review* 9 (1): 71–86.

Ruskin, J. 1903 [1856]. Chapter XII: Of the pathetic fallacy. In *Modern Painters: Volume III*. In *The Works of John Ruskin: Library Edition: Volume III*, J. Ruskin, eds. E.T. Cook and A. Wedderburn, 201–220. London: George Allen.

Salzinger, K. 1979. Ecolinguistics: A radical behavior theory approach to language behavior. In *Psycholinguistic Research: Implications and Applications*, ed. D. Aaronson and R.W. Rieber, 109–129. New York: Lawrence Erlbaum.

Searle, J.R. 1976. A classification of illocutionary acts. *Language in Society* 5 (1): 1–23.

Short, M. 1996. *Exploring the Language of Poems, Plays and Prose*. Harlow: Longman.

Simpson, P. 1993. *Language, Ideology and Point of View*. London and New York: Routledge.

Simpson, P. 2010. Point of view. In *Language and Style: In Honour of Mick Short*, ed. D. McIntyre and B. Busse, 293–310. Basingstoke: Palgrave Macmillan.

Simpson, P. 2014 [2004]. *Stylistics: A Resource Book for Students*, 2nd ed. London and New York: Routledge.

Šklovsky, V. 2012 [1965]. Art as technique. In *Russian Formalist Criticism: Four Essays*, 2nd ed., ed. L.T. Lemon and M.J. Reis, 3–24. Lincoln and London: University of Nebraska Press (first published in Russian in 1917).

Slovic, S., S. Rangarajan, and V. Sarveswaran. 2019. Introduction. In *Routledge Handbook of Ecocriticism and Environmental Communication*, ed. S. Slovic, S. Rangarajan, and V. Sarveswaran, 1–11. London and New York: Routledge.

Soja, E.W. 1996. *Thirdspace: Journeys to Los Angeles and Other Real-and-Imagined Places*. Cambridge, MA: Blackwell.

Sotirova, V., ed. 2016. *The Bloomsbury Companion to Stylistics*. London: Bloomsbury.

Sperber, D., and D. Wilson. 1995 [1986]. *Relevance: Communication and Cognition*, 2nd ed. Oxford: Blackwell.

Steen, G. 2014. Metaphor and style. In *The Cambridge Handbook of Stylistics*, ed. P. Stockwell and S. Whiteley, 315–328. Cambridge: Cambridge University Press.

Steen, G. 2016. Metaphor: Metaphor and style through genre, with illustrations from Carol Ann Duffy's *Rapture*. In *The Bloomsbury Companion to Stylistics*, ed. V. Sotirova, 308–324. London: Bloomsbury.

Steen, G., A.G. Dorst, J.B. Herrmann, A.A. Kaal, T. Krennmayr, and T. Pasma. 2010. *A Method for Linguistic Metaphor Identification: From MIP to MIPVU*. Amsterdam & Philadelphia: John Benjamins.

Steffensen, S.F., and A.F. Fill. 2014. Ecolinguistics: The state of the art and future horizons. *Language Sciences* 41(Part A): 6–25.

Stibbe, A. 2012. *Animals Erased: Discourse, Ecology, and Reconnection with the Natural World*. Middletown, CT: Wesleyan University Press.

Stibbe, A. 2017. *The Stories We Live By: A Free Online Course in Ecolinguistics*. The University of Gloucestershire and The International Ecolinguistics Association. http://storieswliveby.org.uk/. Last accessed April 2022.

Stibbe, A. 2021 [2015]. *Ecolinguistics: Language, Ecology and the Stories We Live By*, 2nd ed. London and New York: Routledge.

Stöckl, H., and S. Molnar. 2018. Eco-advertising: The linguistics and semiotics of green(-washed) persuasion. In *The Routledge Handbook of Ecolinguistics*, ed. A.F. Fill and H. Penz, 261–276. London and New York: Routledge.

Stockwell, P., and S. Whiteley, eds. 2014. *The Cambridge Handbook of Stylistics*. Cambridge: Cambridge University Press.

Talmy, L. 2000. *Toward a Cognitive Semantics: Volume I: Concept Structuring Systems*. Cambridge, MA: The MIT Press.

Tansley, A.G. 1935. The use and abuse of vegetational concepts and terms. *Ecology* 16: 284–307.

Toolan, M. 2014. The theory and philosophy of stylistics. In *The Cambridge Handbook of Stylistics*, ed. P. Stockwell and S. Whiteley, 13–31. Cambridge: Cambridge University Press.

Trampe, W. 2018. Euphemisms for killing animals and for other forms of their use. In *The Routledge Handbook of Ecolinguistics*, ed. A.F. Fill and H. Penz, 325–341. London and New York: Routledge.

United Nations General Assembly (UNGA). 1992. *Report of the United Nations Conference on Environment and Development: Annex I: Rio Declaration on*

Environment and Development. https://www.un.org/en/development/desa/population/migration/generalassembly/docs/globalcompact/A_CONF.151_26_Vol.I_Declaration.pdf. Last accessed April 2022.

United Nations General Assembly (UNGA). 2012. *Resolution Adopted by the General Assembly on 27 July 2012: 66/288. The Future We Want*. http://www.un.org/ga/search/view_doc.asp?symbol=A/RES/66/288&Lang=E. Last accessed April 2022.

United Nations General Assembly (UNGA). 2015. *Resolution Adopted by the General Assembly on 25 September 2015: 70/1. Transforming Our World: The 2030 Agenda for Sustainable Development*. http://www.un.org/ga/search/view_doc.asp?symbol=A/RES/70/1&Lang=E. Last accessed April 2022.

Uspensky, B. 1973. *A Poetics of Composition*, trans. V. Savarin and S. Wittig. Berkeley, CA: University of California Press.

van Peer, W. 1986. *Stylistics and Psychology: Investigations of Foregrounding*. London: Croom Helm.

Vermeulen, K. 2016. The construction of the enemy city in the Hebrew Bible: A case of Thirdspacing. *University of Bucharest Review* 6 (1): 70–83.

Vermeulen, K. 2017a. Home in Biblical and Antwerp city poems: A journey. *Arcadia* 52 (1): 161–182.

Vermeulen, K. 2017b. Save or sack the city: The fate of Jonah's Nineveh from a spatial perspective. *Journal for the Study of the Old Testament* 42 (2): 233–246.

Vermeulen, K. 2020. *Conceptualizing Biblical Cities: A Stylistic Study*. Basingstoke: Palgrave Macmillan.

Virdis, D.F. 2012. Ideological landscapes in G.O. Trevelyan's "An Indian railway": An ecostylistic analysis. In *Ekologiya yazyka na perekrestke nauk: materialy 2-y mezhdunarodnoy nauchnoy konferentsii: v 2-h chastyah: Ch. 1* (*Ecology of Language at the Crossroads of Sciences: Proceedings of the 2nd International Scientific Conference in 2 Volumes: Volume 1*), ed. N.N. Belozerova, 40–47. Tyumen, Russia: Izdatelstvo Tyumenskogo gosudarstvennogo universiteta (Tyumen State University Publishing House).

Virdis, D.F. 2015. Hybrid animals and hybridizing representational strategies in R. Bach's *Jonathan Livingston Seagull*: A functional grammatical investigation. In *Tiere – Texte – Transformationen: Kritische Perspektiven der Human-Animal Studies*, ed. R. Spannring, R. Heuberger, G. Kompatscher, A. Oberprantacher, K. Schachinger, and A. Boucabeille, 213–226. Bielefeld, Germany: Transcript Verlag.

Virdis, D.F. 2016. Mediterranean landscapes as heritage: An ecostylistic analysis of a Sardinian tourism website. In *Mediterranean Heritage in Transit: (Mis-)*

Representations via English, ed. L. Abbamonte and F. Cavaliere, 161–177. Newcastle upon Tyne: Cambridge Scholars Publishing.

Virdis, D.F. 2019. Sexualised landscapes and gentry masculinity in Victorian scenery: An ecostylistic examination of a pornographic novel from the magazine *The Pearl*. *Journal of Literary Semantics* 48 (2): 109–128.

Virdis, D.F. 2021. "She enjoys being stroked", "They are affectionate, lively and interactive boys": An ecostylistic scrutiny of animal agency and alternative discourse in Battersea Dogs & Cats Home Website. In *Language in Place: Stylistic Perspectives on Landscape, Place and Environment*, ed. D.F. Virdis, E. Zurru, and E. Lahey, 189–207. Amsterdam and Philadelphia: John Benjamins.

Virdis, D.F. 2022. Environmental issues in the Victorian era: An ecostylistic examination of metaphor and framing in Ruskin's *The Storm-Cloud of the Nineteenth Century*. *Text & Talk: An Interdisciplinary Journal of Language, Discourse & Communication Studies* 42 (4): 613-632.

Virdis, D.F., ed. Forthcoming. *Ecostylistics: Texts, Theories and Approaches*. Special Issue of *Journal of World Languages*.

Virdis, D.F., and G. Milia. 2021. Exploring feminised landscapes in Victorian erotica: Ecocriticism meets sociology. In *Thinking Out of the Box in Literary and Cultural Studies*, ed. R. Coronato, M. Parlati, and A. Petrina, 111–124. Padua (Italy): Padova University Press.

Virdis, D.F., and M. Cadeddu. Forthcoming. Questioni ambientaliste e paesaggistiche ottocentesche: La natura vista con gli occhi di Alexander von Humboldt e John Ruskin. In *Raccontare il mondo, descrivere la natura: Alexander von Humboldt a 250 anni dalla nascita*, eds. V. Serra and M. Tanca.

Virdis, D.F., E. Zurru, and E. Lahey, eds. 2021a. *Language in Place: Stylistic Perspectives on Landscape, Place and Environment*. Amsterdam and Philadelphia: John Benjamins.

Virdis, D.F., E. Zurru, and E. Lahey. 2021b. Introduction. In *Language in Place: Stylistic Perspectives on Landscape, Place and Environment*, ed. D.F. Virdis, E. Zurru, and E. Lahey, 1–16. Amsterdam and Philadelphia: John Benjamins.

Werth, P. 1994. Extended metaphor: A text-world account. *Language and Literature* 3 (2): 79–103.

Werth, P. 1999. *Text Worlds: Representing Conceptual Space in Discourse*. Harlow: Longman.

Westphal, B. 2011. *Geocriticism: Real and Fictional Spaces*. Basingstoke: Palgrave Macmillan.

Zurru, E. 2012. Islands, rivers, tigers, humans: An ecostylistic analysis of *The Hungry Tide* by Amitav Ghosh. In *Ekologiya yazyka na perekrestke nauk: materialy 2-y mezhdunarodnoy nauchnoy konferentsii: v 2-h chastyah: Ch. 1* (*Ecology of Language at the Crossroads of Sciences: Proceedings of the 2nd International Scientific Conference in 2 Volumes: Volume 1*), ed. N.N. Belozerova, 47–56. Tyumen, Russia: Izdatelstvo Tyumenskogo gosudarstvennogo universiteta (Tyumen State University Publishing House).

Zurru, E. 2016. Exploring the online advertising of the Maltese archipelago's heritage through ecostylistics. In *Mediterranean Heritage in Transit: (Mis-) Representations via English*, ed. L. Abbamonte and F. Cavaliere, 178–204. Newcastle upon Tyne: Cambridge Scholars Publishing.

Zurru, E. 2017. The agency of *The Hungry Tide*: An ecostylistic analysis. In *The Stylistics of Landscapes, the Landscapes of Stylistics*, ed. J. Douthwaite, D.F. Virdis, and E. Zurru, 191–231. Amsterdam and Philadelphia: John Benjamins.

3

Nature

3.1 Definition and Evaluation

The first of the five marker words (Myerson and Rydin 1996, 6; see Sect. 1.1) to be investigated in this book is "nature". The Oxford English Dictionary (OED Online 2022) offers several definitions of this noun. Those most relevant to this book, which are entered under the major semantic division "IV. Senses relating to the material world", are the following:

> **10.a.** The creative and regulative power which is conceived of as operating in the material world and as the immediate cause of its phenomena. Cf. *balance of nature n.* at BALANCE *n.*[1] 13d.
> **10.b.** Usually with capital initial. This power personified as a female being. Frequently as *Dame Nature* or *Mother Nature*.
> **11.a.** The phenomena of the physical world collectively; *esp.* plants, animals, and other features and products of the earth itself, as opposed to humans and human creations.
> (OED Online 2022)

D. F. Virdis, *Ecological Stylistics*,
https://doi.org/10.1007/978-3-031-10658-3_3

In sense 10.a, nature is depicted as a rather abstract entity (noun phrase "The creative and regulative power [...]"), whose existence can only be thought, imagined or hypothesised, but is not fully manifest or demonstrated (lexical verb "is conceived of"). The collocation of the prepositional phrase "of nature" with the noun "balance" in the set noun phrase "balance of nature" implies the existence of "a state of equilibrium in nature produced by the interaction of living organisms; ecological balance" (OED Online 2022). Equilibrium and ecological balance in nature, as well as the interplay of organisms, are beneficial notions (see Sect. 2.1.2) aligning with the ecosophy of this research. As shown in Sect. 1.3, the ecosophy and its norm no. 1: Valuing Living argue that all humans, animals and plants are interconnected; when interconnectedness is stable, it results in their common flourishing and prosperity, as described in norm no. 2: Wellbeing.

As delineated in sense 10.b, the abstract entity "nature" can be embodied by a female figure ("*Dame Nature* or *Mother Nature*"); in fact, in the framework of ecocriticism and ecological discourse, nature is frequently and metaphorically portrayed as a woman (for discussion of and a critical perspective on the Mother Nature metaphor, see Berman 1994). In this respect, Lovelock (2009) was the creator of the Gaia hypothesis, named after the Greek Earth goddess and offered as a scientific model for the planet. In this hypothesis, the earth is personified as Gaia, a living being, and is conceptualised as a powerful physical and biological integrated system: it consists of both biotic and abiotic parts, has a remarkably stable atmosphere and is capable of self-regulation and equilibrium assuring the survival of the system. We humans, in this model, cannot regulate Gaia, but are merely one of her biotic parts, related to and dependent on her completely; what is more, we are irritants or parasites which can easily be shaken off by her. When the Gaia hypothesis was first put forward, our dependence on the planet was a breakthrough idea, but the entire approach was rejected by mainstream science and labelled as feeble and romantic; nevertheless, it later inspired a large number of research projects in ecology and in the related disciplines (Larson 2011, 10, 213–214; Harré et al. 1999, 105; see also Harré et al. 1999, 62–63).

On the contrary, in sense 11.a nature is represented as a concrete entity, more precisely as a collective body or complex whole constituted by animals, plants and natural and physical phenomena of various types. Furthermore, this sense conceptualises nature "as opposed to humans and human creations", i.e. as being disconnected from and in contrast to us humans as a species, to our culture, society and human-made and manufactured objects. That is to say, not only are we humans not described as a part of nature, but a sharp and problematic dichotomy is also forged between what is defined as nature and our human cultural, social, political and economic institutions. Accordingly, when sense 11.a is judged against the ecosophy of this research, the human-nonhuman dualism it evokes is found to convey a destructive discourse (see Sect. 2.1.4). This is because norm no. 1: Valuing Living considers the ecosphere as an interrelated and interdependent whole composed of all human, animal and vegetable organisms, with no discrete or isolated entities. A beneficial rephrasing of this sense would integrate us humans in nature and embrace us among "The phenomena of the physical world collectively", together with "plants, animals"; it would only exclude our artificial "human creations" from those phenomena.

In their study of Greenspeak, Harré et al. (1999, 38–40) identified manifold uses of the noun "nature" and of the adjective "natural", and five antitheses these words are a part of: the natural and the artificial, the organic and the inorganic, the rural and the urban, the wilderness and the peopled, and the natural and the supernatural. As contended by Larson (2011, 113), "we must overcome such thinking and the metaphors that contribute to it in the interests of sustainability. As many have claimed, in various voices, we must embrace a post-Newtonian ecology that integrates humans in nature". We humans must therefore enhance the interconnectedness between nature and us by going beyond our Western hierarchical frame "subject-humans and object-nature", and by adopting the non-hierarchical frame "subject-humans and subject-nature" already adopted by several Indigenous peoples and nations (Larson 2011, 113–114; see also 115–118). This fully aligns with the ecosophy and its norm no. 1: Valuing Living.

The disciplines of ecology and environmental studies or, rather, recently published or reprinted ecology and environment dictionaries and textbooks (see references in the discussion below) do not explicitly define the concept of nature but take it for granted. The only exception among the texts I consulted is the *Bloomsbury Dictionary of Environment & Ecology* (Collin 2011, 143), which features this definition:

> **nature** *noun* all living organisms and the environments in which they live **to go back to nature** to return to a state similar to the one that existed before cultivation ○ *They've let the hillside go back to nature.* → **back-to-nature** (Collin 2011, 143)

In contrast to the meaning of the noun "nature" in sense 11.a of the Oxford English Dictionary, that supplied by the *Bloomsbury Dictionary* is beneficial and in accordance with the ecosophy. "Nature" here does not seem to imply an irreconcilable difference between our human reality and the more-than-human reality, but each living being from both is encompassed in this meaning, as well as their dwellings: this agrees with norm no. 1: Valuing Living. The dictionary entry also defines the idiomatic expression "to go back to nature" and refers to another dictionary entry, that for the adjective "back-to-nature". This adjective is delineated as "relating to a simple self-sufficient way of life with very little use of modern technology" (Collin 2011, 16). This is another beneficial meaning in line with norms no. 2: Wellbeing and no. 5: Environmental Limits: the former maintains that we humans must value basic material and spiritual pleasures, and the latter that we must stop wasting goods, materials and energy.

As previously noted, the *Bloomsbury Dictionary* definition of the noun "nature" is exceptional among the ecology and environment texts referenced in this book, given that it is the only explicit definition of "nature" to be included in them. This evidence is strengthened by the results of Harré et al.'s (1999, 40) study: "We rarely find the concept Nature appearing alone in a Greenspeak document". These authors add that the noun "nature" and the adjective "natural" usually serve as qualifiers of generic notions, as in the noun phrases "natural resource" and "natural cause", or occur in such set phrases as "threats to nature" (Harré et al. 1999, 40–41). Indeed, most of the ecology and environment texts referred to here offer definitions of noun phrases which are near synonyms of "nature" in the contexts of the texts, their subject area and their disciplinary discourse. In addition, the beneficial definition of "nature" examined above is exceptional for at least one other reason: the meanings of its near synonyms as intended in the texts are all destructive. These four noun phrases are listed and defined in the glossary of *Essentials of Ecology* (Miller and Spoolman 2019, G8), a very popular textbook now in its eighth edition:

> **natural capital** Natural resources and natural services that keep us and other species alive and support our economies. See *natural resources, natural services.*
> **natural income** Renewable resources such as plants, animals, and soil provided by natural capital.
> **natural resources** Materials such as air, water, and soil and energy in nature that are essential or useful to humans. See *natural capital.*
> **natural services** Processes of nature, such as purification of air and water and pest control, which support life and human economies. See *natural capital.*
> (Miller and Spoolman 2019, G8)

In sum, "natural capital" consists of "natural resources" plus "natural services", and supplies "natural income". In these noun phrases, the adjective "natural" premodifies the nouns "capital", "income", "resources" and "services", all of which belong to the semantic fields of economics, commerce and administration. These value-laden nouns denote the sum, wealth or stock utilised in further production or to enter into business ("capital"); the money received from employment, business or investments ("income"); stocks, reserves or available assets which are assumed to be there for our human use, domestication, management, exploitation and experimentation, and which can be drawn on when necessary, living beings included ("resources"; see Harré et al. 1999, 40 and Sect. 6.2); assistance or facilities provided to someone to meet their needs or for their use ("services"). These denotations of utility and material benefit are reinforced by a number of words in the definitions of the noun phrases, for example "support our economies", "useful to humans" and "support […] human economies".

The four noun phrases and their meanings hence entail that nature is a set of economic, commercial and administrative resources and services allowing us humans to live and our economies to thrive. The idea of us humans living is backgrounded by the stylistic choices in the text; the idea of more-than-human life is not even present in the text; and the idea of economies thriving is foregrounded. In other words, nature is depicted as having no inherent value or worth in itself; this is confirmed by the definition of the noun phrase "natural resources" in Bailey (2004, 161) and by those of the noun phrases "natural capital" and "natural resource" in Collin (2011, 143) (see also "ecosystem services" in Sect. 5.1).

On the contrary, underlying the ecosophy of this research is the principle, communicated by norm no. 4: Care, that respect and concern are due to all the "materials" catalogued in the definitions above, viz. "plants, animals, and soil" and "air, water, and soil and energy in nature". In line with norm no. 1: Valuing Living, such a careful, respectful and concerned stance must be adopted irrespective of the usefulness of these natural materials and components to our human purposes and profit. Therefore, as anticipated above, the four noun phrases and their definitions are destructive when evaluated against the ecosophy, and must be reworded so that they relay beneficial priorities. In such a rewording, the nouns "capital", "income", "resources" and "services" would not be deployed and, following norm no. 1: Valuing Living, would be replaced by value-neutral words not belonging to the spheres of economics, commerce or administration, like "materials" and "processes" or, in even plainer terms, "nature" and "the natural world". Furthermore, no indications would be given about their being helpful to us humans or their promoting our human economies and interests. Possible rewritings could read as follows:

natural materials Natural components such as air, water, and soil and energy in nature that are essential to our human life and to more-than-human life.

natural processes Processes of nature, such as purification of air and water and ecological balance, which support our human life, more-than-human life and our physical and mental wellbeing.

nature Natural materials and natural processes; among the numerous and varied functions they perform, they keep us humans and other animal and vegetable species alive, and support our physical and mental wellbeing.

In these rewritings, the concept of "nature" is composed of those of "natural materials" and "natural processes", and does not furnish any "natural income". "Natural resources" are now "natural materials". They are not "useful" any longer, but have remained "essential to our human life" and have become essential "to more-than-human life" too; "more-than human life" has been added to the definition of "natural processes" as well. The noun phrase "natural services" has been replaced by "natural processes". The term "pest control", from the definition of "natural services", refers to the regulation of an animal species defined as a pest; as such, it is a case of utilitaristic anthropocentrism, or language use denoting the supposedly harmful aspects of animals, plants and places (Heuberger 2018, 343). As a result, this destructive term has been left out, and the beneficial notion of ecological balance, discussed at the beginning of this section, has been substituted for it. The noun phrase "human economies" has become "our physical and mental wellbeing", both in the definition of "natural processes" and in that of "nature". Anthropocentric "natural capital" is now, simply, ecocentric "nature". The new beneficial definition of this noun acknowledges that nature performs "numerous and varied functions": all of them contribute to the overall ecological balance, to a healthy Gaia and to her healthy guests (for another beneficial alternative, see Sect. 7.2.1 and the idea of "the living processes which ensure food and water for all").

3.2 "Nature" in WWF's Words

3.2.1 Ecostylistic Analysis

The marker word "nature" has now been defined from the lexicological viewpoint of the Oxford English Dictionary (OED Online 2022) and from the naturalistic viewpoint of ecology and environmental studies. How is this marker word employed in the environet (Myerson and Rydin 1996, 7; see Sect. 1.1), in particular in the environmental website and discourse of World Wide Fund for Nature (WWF) (https://wwf.panda.org/) (see Sect. 1.2)? What beneficial ecostylistic strategies are associated with it? Text 1 is drawn from the article "Biocultural diversity: Threatened species, endangered languages", which was posted on WWF's website on 8 June 2014. The article is the opening section of the organisation's report of the same title (Loh and Harmon 2014).

Text 1

[1] It is customary to think of nature and culture as being quite different, belonging to entirely separate domains, one contains items such as butterflies, the Amazon rainforest and photosynthesis, while the other contains items such as Beethoven's piano sonatas, wedding ceremonies or sushi. [2] Yet nature and culture often interpenetrate and overlap. [3] What is wine-making, bee-keeping or gardening: nature or culture? [4] They are undoubtedly human activities, and each has its own culture, but there is a strong element of nature involved. [5] What about varieties of domesticated plants and animals? [6] They are human creations because their genomes have been altered by thousands of generations of selective breeding, and particular breeds may be associated with particular places or peoples, so they are as much a product of culture as of nature. [7] What about landscapes? [8] Is there anywhere left in the world that is entirely natural, untouched by human intervention? [9] The deep sea bed perhaps, and possibly Antarctica; but most landscapes are, to a greater or lesser extent, the product of human culture too. [10] Even the Amazon rainforest is what it is not just because of the natural evolution of its ecosystems, but also because of centuries of human manipulations to those ecosystems. [11] So would it make more sense to think of all the myriad manifestations of nature and culture as expressions of a single concept, a nature-culture nexus?

(© 2014 WWF (https://wwf.panda.org/wwf_news/?222890/Biocultural-Diversity-Threatened-Species-Endangered-Languages). Some rights reserved. Last accessed April 2022)

[225 words]

Underpinning Text 1 is the marker word "nature", which occurs seven times in six different sentences (Sentences 1–4, 6 and 11 [twice]) out of eleven total sentences. As will be discussed below, underlying Text 1 is also the noun "culture", which is present eight times in seven sentences

(Sentences 1–4, 6, 9, 11 [twice]). This is because the text firstly elaborates the hypothesis of a nature-culture opposition (Sentence 1), which is subsequently eliminated and replaced by the concept of "a nature-culture nexus" (Sentence 11). Accordingly, the two nouns frequently appear together in coordinated noun phrases, like "nature and culture" (Sentences 1, 2 and 11), "nature or culture" (Sentence 3), "[a product] of culture as of nature" (Sentence 6) and "nature-culture [nexus]" (Sentence 11).

Sentence 1 and the occurrence of the marker word "nature" in this sentence are founded on the definition of this noun furnished by the Oxford English Dictionary (sense 11.a) and examined in Sect. 3.1, namely "The phenomena of the physical world collectively; *esp.* plants, animals, and other features and products of the earth itself, as opposed to humans and human creations". An antithesis between nature and us humans emerges from this sense: nature is represented as being detached from and opposed to us humans and to the creations of our human culture and experience. This sense is destructive, since it runs counter to the ecosophy of this research and its norm no. 1: Valuing Living: this norm views the ecosphere as a single interconnected system consisting of all humans, animals and plants, with no distinct or separate organisms. As pointed out by Harré et al. (1999, 38–40), this is not an exception in Greenspeak, as various destructive dichotomies are based on the notion of nature. According to Larson (2011, 113), we humans must defeat these dichotomies and adopt a beneficial post-Newtonian ecology connecting us with the natural world.

All the sentences following Sentence 1 entertain Larson's suggestion: they go beyond the Oxford English Dictionary destructive definition of the noun "nature" and try to embrace that provided by the *Bloomsbury Dictionary of Environment & Ecology* (Collin 2011, 143), i.e. "all living organisms and the environments in which they live". Because this definition does instead comply with the ecosophy and its norm no. 1: Valuing Living, it is beneficial: it indicates that nature is composed of all human, animal and vegetable species on the planet, and that there is no separation between our human world and the animal and vegetable world. Inspired by this definition, as anticipated above, Text 1 puts forward the notion of "a nature-culture nexus" (Sentence 11). This is linguistically realised through the stylistic practices presented in the theoretical Sect. 2.2: foregrounding, point of view and metaphor.

With regard to the first of these stylistic practices, Text 1 opens with two finite clauses in Sentence 1: "one [nature] contains items such as butterflies, the Amazon rainforest and photosynthesis, while the other [culture] contains items such as Beethoven's piano sonatas, wedding ceremonies or sushi". Underpinning the clauses is the parallelism "one/the other contains items such as 1, 2 and/or 3", where numerals stand for the parts of the parallel pattern which are varied. In compliance with the "'parallelism rule' for interpretation" (Short 1996, 14; see Sect. 2.2.1), it emerges that the meaning connection made by the various parts is one of contrast: natural "butterflies, the Amazon rainforest and photosynthesis" are opposed to cultural "Beethoven's piano sonatas, wedding ceremonies or sushi". All the remaining sentences in Text 1, however, demonstrate that this nature-culture opposition is only apparent (for the use of another apparent nature-culture opposition in a beneficial ecological text, see Sect. 5.2 on Text 3).

This demonstration is achieved by means of the repetition (Leech and Short 2007, 39–41; Mukařovský 1977; see Sect. 2.2.1) of the nouns "nature" (seven times) and "culture" (eight times) in the entire Text 1, and by means of another parallel pattern in Sentences 3–10. Odd Sentences 3 and 5 ask wh- questions ("What is / What about") based on the opposition; even Sentences 4 and 6 offer answers to those questions ("They are") defeating the opposition. More precisely, Sentences 3 and 5 wonder about the essential character of "wine-making, bee-keeping or gardening" (Sentence 3) and of "varieties of domesticated plants and animals" (Sentence 5). The replies supplied in Sentences 4 and 6 are divided into two parts and founded on another parallelism. The first part of the two sentences claims that these activities and entities are products of culture ("They are undoubtedly human activities, and each has its own culture" in Sentence 4, and "They are human creations because [...]" in Sentence 6). The second part asserts that these are equally products of nature ("there is a strong element of nature involved" in Sentence 4, and "they are as much a product of culture as of nature" in Sentence 6).

Furthermore, Sentence 7 asks a third wh- question ("What about landscapes?") about the nature-culture opposition, followed and glossed by a yes–no question in Sentence 8 ("Is there anywhere […] by human intervention?"). The two questions are answered in the two Sentences 9 and 10, which are against the opposition (for the second yes–no question in Sentence 11, see also below). Here, what is wondered about is the essential character of such landscapes and places as "The deep sea bed", "Antarctica" (Sentence 9) and "the Amazon rainforest" (Sentence 10). Like those in Sentences 4 and 6, the replies in Sentences 9 and 10 are based on a parallel pattern and organised into two parts; unlike that in Sentences 4 and 6, this pattern first mentions nature, then culture. Hence, the first part of Sentences 9 and 10 maintains that "The deep sea bed perhaps, and possibly Antarctica" and "the Amazon rainforest" are products of nature (via ellipsis, Sentence 9 anaphorically refers to Sentence 8 and to the adjective phrases "entirely natural, untouched by human intervention"; see also "the natural evolution of its ecosystems" in Sentence 10). The second part contends that "most landscapes" and "the Amazon rainforest" are products of culture as well ("the product of human culture too" in Sentence 9, and "centuries of human manipulations to those ecosystems" in Sentence 10).

The question–answer structure in Sentences 3–10 constitutes an instance of discoursal deviation (Short 1996, 37–43; see Sect. 2.2.1). A prototypical article shows an addresser (the author) conveying a message (the article) to an addressee (their reader). On the contrary, the message relayed in Text 1 appears as a dialogue between two participants: an unlearned one asking questions and a learned one supplying answers. Therefore, the question–answer parallelism furnishes Text 1 with a regular rhythm, which makes the text easier and more pleasant to read; the parallel pattern also sets down a detailed list of examples and reasons to reject the nature-culture opposition, thus encouraging the addressee to do the same (for the use of another list in a beneficial ecological text, see Sect. 4.2 on Text 2).

Text 1 communicates its two addressers' point of view, in particular their psychological point of view, through epistemic and perception modality. In other words, the text suggests the addressers' knowledge, beliefs and cognition, and whether the addressers are certain of the veracity of the propositions they write; in the case of perception modality, underlying the addressers' knowledge, beliefs and cognition are visual and other powers of perception (for point of view and modality, see Simpson 1993, 43–47; summarised in Neary 2014, 181–182, and in Sect. 2.2.2). In accordance with the Hallidayan view of modality (Halliday 2014, 176–183), degrees of usuality and oftenness ("sometimes/usually/always") are also incorporated into the epistemic modal system and, as a result, are investigated here. In order to have a more complete and accurate scrutiny, degree adjuncts of intensity (e.g. "totally", "almost", "scarcely") (Halliday 2014, 187–189) are also comprised in epistemic modality and explored. Epistemic modality is signalled in Text 1 by twelve modal operators (eleven words and the phrase "to a greater or lesser extent"), and perception modality is conveyed by one modal operator, the string "make more sense" (with the meaning of "be more intelligible"):

> [1] customary; quite; entirely; [2] often; [4] undoubtedly; [6] may; [8] entirely; [9] perhaps; possibly; most; to a greater or lesser extent; [11] would; make more sense [= be more intelligible]

In Text 1, what propositions are psychologically evaluated by the addressers via these thirteen linguistic indicators of epistemic and perception modality? And what level of addresser commitment to these propositions is relayed? Strong commitment is employed to characterise nature and culture through the adjective phrase "quite different" and the noun phrase "entirely separate domains" (Sentence 1), and to controvert the nature-culture opposition by implying that no place on the planet is "entirely natural" (Sentence 8). In addition, strong commitment, realised by the adjective "more" in the string "make more sense" (Sentence 11), is utilised to express that it is more correct and fact-based to think in terms of a nature-culture nexus. By means of medium commitment, Text 1 communicates that, instead, thinking in terms of the nature-culture opposition is "customary", viz. usually, but not always, the case (Sentence 1). Medium commitment also signals that "most landscapes" (not all) are human creations "to a greater or lesser extent" (not to a full extent) (Sentence 9). This is because, as suggested via the weak-commitment modal adverbs "perhaps" and "possibly", "The deep sea bed" and "Antarctica" are very likely not to be human products (Sentence 9).

In point of view theory, the epistemic and perception modal systems, particularly their lower values, underpin negative modal shading, namely the addresser's uncertainty as to the truthfulness of their assertions (for modal shading, see Simpson 1993, 50–69; summarised in Neary 2014, 183–187, and in Sect. 2.2.2). Nevertheless, most of the thirteen epistemic and perception modal operators in Text 1 indicate the addressers' medium to strong commitment to their propositions; consequently, Text 1 does not suggest negative modal shading. Moreover, this degree of commitment is reinforced by unmodalised categorical assertions, which evoke the highest possible value of addresser commitment, in four sentences and in various clauses: the two finite clauses with the parallelism in Sentence 1; the questions in Sentences 3, 5 and 7; most of Sentence 4 (about wine-making, bee-keeping and gardening); most of Sentence 6 (about varieties of domesticated plants and animals); Sentence 10 (about the Amazon rainforest). Such assertions constitute neutral modal shading; that is, propositions are not psychologically evaluated. In short, the propositions in Text 1 in favour of the nature-culture nexus and against the nature-culture opposition are assessed by the addressers as being accurate, if not factual; hence, they powerfully urge the addressee to share the addressers' psychological point of view on nature.

Urgency is directly conveyed by Sentence 11 ("So would [...] nature-culture nexus?"), the only question not to receive an explicit answer in Text 1; it thereby realises internal deviation from the question–answer parallelism in Sentences 3–10. Sentence 11 is a request for the addressee's opinion on the nature-culture nexus: "So would it make more sense [...]" can be rephrased as "So do you think it makes more sense [...]" (Halliday 2014, 177). After the list of answers provided by the parallel pattern, after the addressers' epistemic modal expressions of commitment to those answers and after their unmodalised categorial assertions, the addressee's own answer to this final question cannot but be "Yes, it would!". In traditional terms, this is a rhetorical question implicitly assuming a preferred (usually negative, but here positive) answer; its communicative function is not to get an answer or gain information, but to produce a foregrounding effect persuading or emotionally affecting the addressee (Wales 2011, 370–371; for the use of another rhetorical question in a beneficial ecological text, see Sect. 4.2 on Text 2).

Two unmodalised categorical assertions in Sentence 1 (the two parallel finite clauses "one [nature] contains items such as [...], while the other [culture] contains items such as [...]") include two conceptual metaphors (Lakoff and Johnson 2003; see Sect. 2.2.3). In line with Steen's (2016; see Sect. 2.2.3) three-dimensional model of metaphor, the two conceptual metaphors can be scrutinised as follows:

1. At the level of expression, or metaphor in language, the two conceptual metaphors are triggered by the predicator-verb phrase "contains", occurring in the two clauses. Its two subjects are the noun phrases "one [nature]" and "the other [culture]", and its two direct objects are the noun phrases "items such as [...]" (twice).
2. At the level of conceptualisation, or metaphor in thought, the two subjects "one [nature]" and "the other [culture]" constitute the targets NATURE and CULTURE, and the predicator "contains" the source CONTAINER: this results in the two conceptual metaphors NATURE IS A CONTAINER and CULTURE IS A CONTAINER. The main feature of the source domain mapping to the two target domains is the fact that a CONTAINER can hold objects. Actually, the items mentioned in the two direct objects ("butterflies, the Amazon rainforest and photosynthesis", and "Beethoven's piano sonatas, wedding ceremonies or sushi") offer samples of the entities encompassed in the two figurative containers.
3. At the level of communication, or metaphor in interaction, NATURE IS A CONTAINER and CULTURE IS A CONTAINER are creative metaphors with a deliberate figurative function. In fact, neither is listed in the three databases Master Metaphor List (Lakoff and Cognitive Linguistics Group 1991), Metalude (Goatly and LLE Project 2002–2005) and MetaNet Metaphor Wiki (David et al. 2016, 2018).

The fact that both nature and culture are conceptualised as containers attributes to them, by means of cross-domain mapping, the features of these objects, viz. receptacles designed to contain, store or transport some particular articles or goods. The most relevant characteristic of containers mapping to nature and culture in the context of Sentence 1 is their being airtight and hermetically sealed, so that the articles in them are held securely and cannot leave the container or get mixed with articles of other types. However, in Sentence 2 ("Yet nature and culture often interpenetrate and overlap"), a new conceptual metaphor contradicts this characteristic, along with the nature-culture opposition:

1. At the level of expression, or metaphor in language, the new conceptual metaphor is prompted by the coordinated predicator-verb phrase "interpenetrate and overlap". In the previous conceptual metaphors, the two independent noun phrases "one [nature]" and "the other [culture]" realised two subjects, while in the new conceptual metaphor the coordinated noun phrase "nature and culture" realises one subject.
2. At the level of conceptualisation, or metaphor in thought, the coordinated subject "nature and culture" constitutes the target NATURE AND CULTURE, and the coordinated predicator "interpenetrate and overlap" supplies the traits of the source, which are those of COMMUNICATING VESSELS. The source represents the target as two containers being able to gain entrance or access to each other, with the contents of one container partly coinciding with those of the other. Accordingly, the new conceptual metaphor reconceptualises the two independent targets NATURE and CULTURE, and the two conceptual metaphors NATURE IS A CONTAINER and CULTURE IS A CONTAINER become the new conceptual metaphor NATURE AND CULTURE ARE COMMUNICATING VESSELS. The various contents of the two vessels are free to move from one to the other, to combine and to achieve a stable balance; that is to say, as Text 1 puts it, to create the nature-culture nexus.

3. At the level of communication, or metaphor in interaction, NATURE AND CULTURE ARE COMMUNICATING VESSELS has a deliberate figurative function. It is, therefore, a creative metaphor not figuring in the three metaphor databases referenced above.

While the three conceptual metaphors explored so far are creative, Text 1 comprises a metaphor which is conventionalised and lexicalised (Goatly and LLE Project 2002–2005), but equally worth analysing in this book: the noun "manipulations", employed in Sentence 10 ("human manipulations to those ecosystems [the Amazon rainforest]"). In the Oxford English Dictionary (OED Online 2022), the entry "manipulation" has a literal meaning in sense 3.a: "The action or an act of manipulating something; handling; dexterity". As a result, the manipulations performed by the Amazon nations on the ecosystems of the Amazon rainforest are framed as physical handling of their natural materials and processes. These manipulations and this handling by the Amazon nations are positively value-laden, for these peoples have lived in the Amazon rainforest for centuries, or even thousands of years, and have managed to maintain ecological stability in the entire ecosystem (for the issue of Indigenous nations and peoples in beneficial ecological texts, see also Text 2 and Text 4 and Sects. 4.2 and 6.2).

Nevertheless, the Oxford English Dictionary entry "manipulation", in sense 4, has a more figurative, and negatively value-laden, meaning: "The action or an act of managing or directing a person, etc., esp. in a skilful manner; the exercise of subtle, underhand, or devious influence or control over a person, organization, etc.; interference, tampering". This negative denotation is also present in the three metaphor databases:

1. In Master Metaphor List, the conceptual metaphor [PSYCHOLOGICAL] MANIPULATION IS PHYSICAL MANIPULATION has such linguistic realisations as "He twisted everyone's emotions"; its subcase BEING EASILY [PSYCHOLOGICALLY] MANIPULATED IS BEING EASILY PHYSICALLY MANIPULATED has linguistic realisations like "She was silly putty in his hands".
2. In Metalude, the lexical item "manipulate", with the literal meaning "move and press with the hands", can be found in the root analogy

CONTROL IS HANDLE/OWN, with the metaphorical meaning "control unfairly or dishonestly"; an example of this is "Enron were manipulating the statistics to suit themselves".

3. In MetaNet, the conceptual metaphor CONTROL IS OBJECT MANIPULATION is illustrated by the example text "He's putty in her hands". The source frame role "actor" maps to the corresponding target frame role "controlling entity"; and the source frame role "object" maps to the corresponding target frame role "dependent entity" (for frames, see Tannen 1993).

Against this negatively value-laden background, the noun phrase "human manipulations to those ecosystems [the Amazon rainforest]" in Text 1 may bring to mind such radical interference and tampering as deforestation due to human settlement, speculative land development, crop cultivation, cattle pasture and oil drilling. Following the MetaNet frame roles, the Amazon nature and ecosystems are represented as passive objects dependent on us humans, and we humans as active actors controlling nature and those ecosystems. These manipulations are not undertaken, nor are these frame roles assigned, by the Amazon nations, but by us humans from the Global North. In Metalude's words, such manipulations are conceptualised as unfair and dishonest control of nature; and, in the Oxford English Dictionary words, as subtle, underhand and devious influence over the environment. Consequently, the human manipulations of the Amazon rainforest ecosystems pursued by the Amazon nations, along with their mutual connection with those ecosystems, are a prime instance of the nature-culture nexus. Conversely, the human manipulations carried out on those ecosystems by us humans from the Global North are an example of environmentally, socially and morally unacceptable behaviour, the actions of an unecological social system still believing in the nature-culture opposition.

3.2.2 Summary and Evaluation

After the ecostylistic examination of Text 1, taken from WWF's website, containing and defining the marker word "nature", and the study of the stylistic techniques of foregrounding, point of view and metaphor appearing in the text, Text 1 and these techniques will be summarised and appraised. To be more precise, this section judges the value-system of Text 1, draws a comparison with the ecosophy of this research and its norms, and aims to show whether the text and its techniques are indeed beneficial.

As anticipated in Sect. 3.2.1, the idea of the nature-culture nexus in Text 1 is based on the beneficial definition of "nature" furnished by the *Bloomsbury Dictionary of Environment & Ecology* (Collin 2011, 143), which includes both our human world and the more-than-human world in nature. Therefore, the nature-culture nexus idea rebuffs the destructive definition of this marker word provided by the Oxford English Dictionary (OED Online 2022), which excludes our human world from nature. This idea is also inspired by Larson's (2011, 113) suggestion that we humans "embrace a post-Newtonian ecology that integrates humans in nature". The concept of the nature-culture nexus aligns with the ecosophy and its norm no. 1: Valuing Living. In Text 1, culture is not in contrast to nature, since underpinning our human cultural artefacts, procedures and customs is nature. Hence, our human lives have the same intrinsic value as those of all the other animal and vegetable species on earth, and all life, both human and more-than-human, must be considered very highly. The fact that human culture is intimately linked to and embedded in nonhuman organisms implies that these organisms are living entities equal to us humans, as such deserving empathy, regret and gratitude when we kill them or affect them adversely: this is norm no. 4: Care in the ecosophy. Accordingly, the nature-culture nexus is a beneficial notion to be promoted among the general public. The ecosophy and its norms no. 1: Valuing Living and no. 4: Care are relayed and highlighted in Text 1 through the stylistic mechanisms of foregrounding, point of view and metaphor; in a number of sentences, two or three mechanisms co-occur to mark critical points in the text and to make a stronger impression on the addressee.

The foregrounding device of parallelism is first utilised in Sentence 1 to introduce the destructive idea of the nature-culture opposition ("one [nature] contains items such as butterflies, the Amazon rainforest and photosynthesis, while the other [culture] contains items such as Beethoven's piano sonatas, wedding ceremonies or sushi"). Subsequently, in Sentences 3–10, a longer and more complex parallel pattern is deployed to extensively counter the opposition with the beneficial proposal of the nature-culture nexus: Sentences 3, 5, 7 and 8 ask questions about the opposition and the essential character of allegedly cultural or natural entities ("What is [...]", "What about [...]" [twice], "Is there [...]"), and Sentences 4, 6, 9 and 10 give answers against the opposition. This second parallelism co-occurs with and is strengthened by the discoursal deviation consisting of the question–answer sequence resembling a dialogue between two people: an uneducated one who asks questions and an educated one who answers them. The stylistic function of parallelism and of discoursal deviation in Text 1 is therefore to emphasise and make more persuasive the nonexistence of the destructive nature-culture opposition and the existence of the beneficial nature-culture nexus; they are thus utilised to recall the ecosophy and its norms no. 1: Valuing Living and no. 4: Care.

Text 1 communicates a beneficial value-system in accordance with the ecosophy also by means of point of view, specifically psychological point of view expressed by epistemic modality (twelve modal operators in Sentences 1–11), perception modality (one modal operator, the string "make more sense", in Sentence 11) and lack of modality (in the four Sentences 3, 5, 7 and 10). Throughout Text 1, medium values (e.g. "customary", "most", "to a greater or lesser extent") to high values ("quite", "entirely", "make more sense") of epistemic and perception modality (with no negative modal shading) are employed to signal the addressers' medium to strong commitment to the factuality of the beneficial nature-culture nexus and, as a result, to norms no. 1: Valuing Living and no. 4: Care it agrees with. In addition, four unmodalised categorical assertions (with neutral modal shading) confirm the factuality of the nature-culture nexus and the norms and increase this factuality to the highest value. In the concluding Sentence 11, which features the rhetorical question "So would it make more sense [...]", point of view and modality co-occur with internal deviation from the question–answer parallel pattern present in Sentences 3–10. In Sentence 11, the addressee of Text 1 is directly asked a question and invited to express their opinion on the nature-culture nexus; all the foregrounding and point of view strategies in the text lead them to approve of it wholeheartedly. Consequently, actively engaging the addressee in the text and in its meaning-making processes is shown to be a successful stylistic practice to suggest a beneficial ecosophy valuing all life and caring for all animals and plants.

A sentence in Text 1 that is key to attracting the addressee's attention is the opening Sentence 1, where instances of all the stylistic techniques investigated in this book co-occur—parallelism, psychological point of view and metaphor. However, in the two finite clauses "one [nature] contains items such as [...], while the other [culture] contains items such as [...]", Sentence 1 does not evoke the beneficial value-system founded on the concept of the nature-culture nexus; it evokes the destructive value-system founded on the concept of the nature-culture opposition. This is attained through the two creative metaphors NATURE IS A CONTAINER and CULTURE IS A CONTAINER, which run counter to the ecosophy and its norms no. 1: Valuing Living and no. 4: Care, and are hence destructive. Nevertheless, Sentence 1 and the two destructive metaphors have the function of presenting the opening thesis which will be controverted in all the subsequent sentences in Text 1, starting from Sentence 2 ("Yet nature and culture often interpenetrate and overlap"). This sentence introduces into the text the creative metaphor NATURE AND CULTURE ARE COMMUNICATING VESSELS, which underlies the ideas in the following sentences climaxing with the notion of the nature-culture nexus. This creative metaphor represents nature and culture in an innovative and, above all, beneficial way aligning with the ecosophy and its norms no. 1 and no. 4; therefore, it is a stylistic mechanism which must be advertised.

Another stylistic mechanism in Text 1 is the use of the noun "manipulations" in the noun phrase "human manipulations to those ecosystems [the Amazon rainforest]" (Sentence 10). In the text, the noun has the literal meaning defined by the Oxford English Dictionary sense 3.a and refers to the Amazon nations' handling and mutual connection with the ecosystems of the Amazon rainforest. Unlike the "actor" and "object" frame roles from MetaNet, the noun, in its literal meaning and in the context of Text 1, does not conceptualise us humans as controlling actors having malevolent power over and a harmful impact on nature; the noun does not conceptualise nature as an object dependent on us either. Accordingly, the literal use of "manipulations" and the ecological ideas and behaviours it conveys are beneficial and in accordance with the ecosophy. Not only with norms no. 1: Valuing Living and no. 4: Care, but also with norm no. 2: Wellbeing, because the interpenetration between the Amazon nations and the Amazon rainforest has been stable and balanced for hundreds of years: this has resulted in the prosperity and physical and psychological health of all the humans, animals and plants in those ecosystems.

As recorded in the Oxford English Dictionary sense 4 and in the three metaphor databases referenced in this book, the noun "manipulations" also has a conventionalised and lexicalised metaphorical meaning of interference and tampering. This negatively value-laden metaphorical meaning is not found in Text 1; nevertheless, it may suggest itself to the addressee when they read about the Amazon nations, the Amazon rainforest and their ecosystems. Consequently, the noun phrase "human manipulations to those ecosystems [the Amazon rainforest]" could be reworded as "human interconnections with those ecosystems [the Amazon rainforest]": the new noun "interconnections" would not activate the negative metaphorical meaning of interference and tampering, and would not lead the addressee to mistakenly believe that the Amazon nations are involved in these actions. In its negative metaphorical meaning, the noun "manipulations" delineates, in a critical and realistic manner, the present state of affairs and the relations between the human and the more-than-human world in a key area of the earth. Therefore, the metaphorical use of "manipulations" recalls the ecosophy and its norms no. 3: Now and the Future, no. 5: Environmental Limits and no. 6: Social Justice.

Firstly, norm no. 3: Now and the Future states that the ecosophy has an unrestricted time range, given that it takes into consideration the ability of all animal and vegetable species to live well in the present and in the future alike. If our human manipulations of the Amazon rainforest and of the entire planet continue to be dramatic, however, there will be no future; as a result, serious steps must be taken now to preserve the Amazon rainforest and all the ecosystems on earth in the future. Secondly, in norm no. 5: Environmental Limits, it is argued that we humans are damaging the Amazon rainforest and a large number of ecosystems by our excessive use of their natural materials and processes and by the waste we generate. If our human manipulations continue to be substantial, the environmental limits of the planet will be overstepped and its ecological harmony permanently damaged. Thirdly, norm no. 6: Social Justice contends that the most socially vulnerable peoples on earth are the poor, powerless and often displaced populations from the Global South living in the most ecologically vulnerable ecosystems; they are frequently the victims of the phenomenon of slow violence (Nixon 2011), namely such dangers as the deforestation of the Amazon rainforest and all its causes mentioned in Sect. 3.2.1. If our human manipulations continue to be destructive, and if those of us from the Global North do not change our social and ecological lifestyles profoundly, the Amazon nations and ecosystems, as well as several weak peoples, ecosystems and sustainable ways of life worldwide, will be lost forever. As emerges from the ecostylistic scrutiny of Text 1 and from its comparison with the ecosophy of this research, thinking in terms of the nature-culture nexus, as recommended by WWF's environmental website, and being inspired by the Amazon nations' beneficial lifestyle may contribute to bringing about a new mindset and a new ecological agenda.

References

Bailey, J., ed. 2004. *The Facts on File Dictionary of Ecology and the Environment*. New York: Market House Books.

Berman, T. 1994. The rape of Mother Nature? Women in the language of environmental discourse. *The Trumpeter* 11 (4): 173–178.

Collin, P.H. 2011 [1985]. *Dictionary of Environment & Ecology*, 5th edition. London: Bloomsbury [electronic edition].

David, O., G. Lakoff et al. 2016–2018. *MetaNet*. Berkeley: University of California. https://metaphor.icsi.berkeley.edu/pub/en. Last accessed April 2022.

Goatly, A., and LLE Project. 2002–2005. *Metalude*. Hong Kong: Lingnan University. http://www.ln.edu.hk/lle/cwd/project01/web/home.html. Last accessed April 2022.

Halliday, M.A.K. 2014 [1985]. *Halliday's Introduction to Functional Grammar*, revised by C.M.I.M. Matthiessen, 4th edition. London and New York: Routledge.

Harré, R., J. Brockmeier, and P. Mühlhäuser. 1999. *Greenspeak: A Study of Environmental Discourse*. London: Sage.

Heuberger, R. 2018. Overcoming anthropocentrism with anthropomorphic and physiocentric uses of language? In *The Routledge Handbook of Ecolinguistics*, ed. A.F. Fill and H. Penz, 342–354. London and New York: Routledge.

Lakoff, G., and Cognitive Linguistics Group. 1991. *Master Metaphor List*. Berkeley: University of California. http://araw.mede.uic.edu/~alansz/metaphor/METAPHORLIST.pdf. Last accessed April 2022.

Lakoff, G., and M. Johnson. 2003 [1980]. *Metaphors We Live By: With a New Afterword*, 2nd edition. Chicago: University of Chicago Press.

Larson, B. 2011. *Metaphors for Environmental Sustainability: Redefining our Relationship with Nature*. New Haven, CT: Yale University Press.

Leech, G., and M. Short. 2007 [1981]. *Style in Fiction: A Linguistic Introduction to English Fictional Prose*, 2nd edition. Harlow: Longman.

Loh, J., and D. Harmon. 2014. *Biocultural Diversity: Threatened Species, Endangered Languages*. Zeist, The Netherlands: WWF Netherlands.

Lovelock, J. 2009. *The Vanishing Face of Gaia: A Final Warning*. New York: Basic Books.

Miller, G.T., and S.E. Spoolman. 2019 [2007]. *Essentials of Ecology*, 8th edition. Belmont, CA: Cengage.

Mukařovský, J. 1977. On poetic language. In *The Word and Verbal Art: Selected Essays*, ed. J. Burbank and P. Steiner, 1–64. New Haven, CT and London: Yale University Press.

Myerson, G., and Y. Rydin. 1996. *The Language of Environment: A New Rhetoric*. London and New York: Routledge.

Neary, C. 2014. Stylistics, point of view and modality. In *The Routledge Handbook of Stylistics*, ed. M. Burke, 175–190. London and New York: Routledge.

Nixon, R. 2011. *Slow Violence and the Environmentalism of the Poor*. Cambridge, MA: Harvard University Press.

Oxford English Dictionary (OED) Online. 2022. Oxford: Oxford University Press. https://www.oed.com/. Last accessed April 2022.

Short, M. 1996. *Exploring the Language of Poems, Plays and Prose*. Harlow: Longman.

Simpson, P. 1993. *Language, Ideology and Point of View*. London and New York: Routledge.
Steen, G. 2016. Metaphor: Metaphor and style through genre, with illustrations from Carol Ann Duffy's *Rapture*. In *The Bloomsbury Companion to Stylistics*, ed. V. Sotirova, 308–324. London: Bloomsbury.
Tannen, D., ed. 1993. *Framing in Discourse*. Oxford: Oxford University Press.
Wales, K. 2011 [1990]. *A Dictionary of Stylistics*, 3rd edition. London and New York: Routledge.

4

Environment

4.1 Definition and Evaluation

"Environment" is the second of the five marker words (Myerson and Rydin 1996, 6; see Sect. 1.1) to be scrutinised in this book. Of the senses of this noun provided by the Oxford English Dictionary (OED Online 2022), the most pertinent to this book are the following:

> **2.a.** The area surrounding a place or thing; the environs, surroundings, or physical context. Also *figurative*.
> **2.b.** The physical surroundings or conditions in which a person or other organism lives, develops, etc., or in which a thing exists; the external conditions in general affecting the life, existence, or properties of an organism or object.
> **2.c.** With modifying word: a particular set of surroundings or conditions which something or someone exists in or interacts with [e.g., semi-desert, marine, acid environment].
> **2.d.** Frequently with *the*. The natural world or physical surroundings in general, either as a whole or within a particular geographical area, esp. as affected by human activity.
> (OED Online 2022)

D. F. Virdis, *Ecological Stylistics*,
https://doi.org/10.1007/978-3-031-10658-3_4

A number of stylistic characteristics in these definitions stand out in the context of this book. In sense 2.b, the coordinating conjunction "or" connects the noun phrases "a person" and "other organism", thus equating them; moreover, the adjective "other" triggers the lexical presupposition that people are organisms like animals and plants. These characteristics make us humans, our lives, development and needs equal to those of nonhuman organisms. In sense 2.d, the words "esp. as affected by human activity" openly acknowledge the fact that our human occupations influence and make a material impression on the environment. In its norms no. 1: Valuing Living and no. 2: Wellbeing, the ecosophy of this research (see Sect. 1.3) contains the concept of valuing the lives and wellbeing of all humans, animals and plants; the ecosophy also contains the concept of environmental limits challenged by us humans in its norm no. 5: Environmental Limits. As a result, the stylistic aspects investigated above are beneficial (see Sect. 2.1.2) when compared with the ecosophy.

Nevertheless, the definitions also seem to feature a remarkable destructive aspect (see Sect. 2.1.4). Such noun phrases as "a person or other organism", "an organism or object" (sense 2.b) and "something or someone" (sense 2.c) apparently imply that the environment is an entity separate from and alien to the human, animal and vegetable species living and developing there. As contended by the biologist Brendon Larson (2011, 121), the meaning of the noun "environment", originally "that which surrounds", has changed to mirror the dichotomy between nature and culture: "this perspective can be taken only by a disengaged humanity, perhaps by the omniscient gaze of objective science. […] This gaze belies an ontology in which the environment is external to us rather than one in which we are embedded, a life world".

A definition of the noun "environment" figures in almost all the recent ecology and environment dictionaries and textbooks in their latest editions or reprints I referred to. The longest and most complex definition can be found in Routledge's *The Environment Dictionary* (Kemp 2002, 127):

> A combination of the various physical and biological **elements** that affect the life of an organism. Although it is common to refer to "the" environment, there are in fact many environments, all capable of change in time

> and place, but all intimately linked and in combination constituting the whole earth/atmosphere system. They vary in scale from microscopic to global and may be subdivided according to their attributes. The **aquatic environment**, for example, is that of rivers, lakes and oceans, the **terrestrial environment** that of the land surface. The term "built environment" has been applied to areas, such as cities, created by human activity. The human element has a dominant role in modern environmental studies, and as a result environmental science includes not only the traditional sciences such as chemistry, physics and biology, but also engineering, economics, sociology, politics and law. The study of the environment is thus very much interdisciplinary in nature. (Kemp 2002, 127)

Several pieces of information here are meaningful to this book. Both the elements impacting on an organism's life and the diverse environments composing the planet are portrayed as being closely combined. This involves a joint action and a common purpose, and assumes the existence of balance, stability and harmony between the elements and the environments within the wider system. In addition, and more importantly, it also assumes that such a balance must not be upset, disturbed or lost, even on an apparently trivial microscopic scale, in order not to endanger not only the lives of the single organisms, but also the life of the entire earth. Out of the 152 words the definition consists of, 65 (more than 40%), constituting its second half, are devoted to the effect of us humans and our actions on the environment and, consequently, to our role in contemporary environmental studies. This role is called "dominant", a trait proven by the list of the five "non-traditional" social sciences and humanities fields required to fully understand and explore it. The list also communicates that environmental studies is an interdisciplinary subject and research area, hence more articulate than others.

The ecosophy of this research, in its norm no. 1: Valuing Living, states that the ecosphere is composed of all human and more-than-human life dwelling there, and champions flourishing and physical and mental health for all species in its norm no. 2: Wellbeing. Therefore, the fact that the definition of the noun "environment" under examination gives prominence to the notion of deeply interconnected elements and environments and to the resulting notion of harmony between them is beneficial and in accordance with the ecosophy. It would have been even more beneficial if it had made explicit the assumption that this harmony must be preserved so as not to threaten life on the planet in the years to come.

Furthermore, the ecosophy comprises the idea of environmental limits in norms no. 3: Now and the Future and no. 5: Environmental Limits: we humans and our consumerist society and culture have damaged the ability of the environment to support life in the near future. As previously shown, the second half of the definition of "environment" is dedicated to our role in environmental research and to our various relations with nature from different disciplinary standpoints. The text-type of the dictionary is supposed to offer seemingly unbiased and value-neutral details. Accordingly, although our role is termed "dominant", no reference is made to the environmental limits strained by our way of life and authoritative role (against norms no. 3 and no. 5). No reference is made to the subordination of animal and vegetable organisms implied by this way of life and this role, either, while all species must be given loving and proper attention, reverence and concern (against norms no. 1: Valuing Living and no. 4: Care). That is to say, the dictionary does not overtly observe that our effect on the environment is often adverse and devastating, never mind lasting and far-reaching.

This is the stylistic and ecolinguistic mechanism of erasure, or "A story in people's minds that an area of life is unimportant or unworthy of attention" (Stibbe 2021, 223; see also Sect. 2.1.4). The use of erasure makes the second half of the definition destructive and not in line with the ecosophy. This destructive section would have been beneficial if the opposing strategy of salience (Stibbe 2021, 227) had been employed, namely if the text had represented our influence on the natural world as "A story in people's minds that an area of life is important or worthy of attention" by openly dealing with our deleterious role. The same issue—the use of the practice of erasure, rather than salience, to play down our impact on the earth—also arises in a number of definitions given by other ecology and environment dictionaries and textbooks. The cases most related to the one above are the noun phrases "environmental degradation", "environmental science" and "environmental scientist" in Miller and Spoolman (2019, G4); "environmental science" in Smith and Smith (2015, 661); "environmental hazard", "environmental load" and "environmentalism" in Bailey (2004, 85); "environmental damage", "environmental degradation", "environmental disorder", "environmental impact" and "environmental science" in Collin (2011, 74).

4.2 "Environment" in Greenpeace's Words

4.2.1 Ecostylistic Analysis

In the previous section, the Oxford English Dictionary (OED Online 2022) and recent ecology and environment dictionaries and textbooks were referenced to offer a lexicological and naturalistic definition of the marker word "environment". This section will now explore the use of this marker word in the environet (Myerson and Rydin 1996, 7; see Sect. 1.1), more precisely in the ecological discourse of Greenpeace International as conveyed by its website (https://www.greenpeace.org/international/) (see Sect. 1.2); in practice, the section will analyse the beneficial ecostylistic strategies connected with the use of this word. Text 2 is drawn from the story (a website category) "Earth to humans ...", which was written by the content editor Therese Salvador, and published on Greenpeace's website on 23 April 2021. Text 2 is the concluding section in the story and is headed by Sentence 1.

Text 2

[1] We Must Restore Our Earth

[2] Mother Earth is estimated to be around 4.5 billion years old. [3] She is weary, disheveled and drained. [4] Given what she's been through, she feels cranky, frustrated and tired of humans. [5] Who can blame her?

[6] We've always been needy and greedy. [7] We keep depleting Earth's natural resources without remorse or restitution. [8] In fact, we've been using up Earth's resources at an alarming rate. [9] Currently, we are using up 1.6 times more than what the Earth can regenerate in a year.

[10] [10a] Though she may be resilient, [10b] we need to give Mother Earth a break—we must offer her a lifeline. [11] We need to stop treating the environment as a commodity and as an infinite resource because doing so would be fatal.

[12] With this "new normal", we must learn to reconnect with nature and consider the environment in everything we do. [13] If there's anything to learn from the pandemic, it is that governments need to shift their priorities and invest on [sic] human and planetary health to make the world resilient to future shocks.

[14] We need to decarbonise economies and shift to renewable energy systems to mitigate global warming. [15] Efforts to protect at least 30% of the world's oceans by 2030 must be supported to save marine biodiversity and to help save the climate. [16] Forests that have stood the test of time and teeming with life must be preserved and protected from corporate interests that want to profit from it [sic]. [17] Indigenous Peoples, with their ancient wisdom of the natural world[,] must be empowered and recognised.

[18] She may be old and complex. [19] But undoubtedly, Earth is a fine place and worth fighting for.

(*Source* Greenpeace International (https://www.greenpeace.org/international/story/47338/earth-to-humans/). Author: Therese Salvador. Date: 23 April 2021. Last accessed April 2022)

[273 words]

In Text 2, the marker word "environment" can be found twice, in two consecutive sentences: Sentence 11 ("We need to stop treating the environment as a commodity and as an infinite resource because doing so would be fatal") and Sentence 12 ("With this 'new normal', we must learn to reconnect with nature and consider the environment in everything we do"). The two sentences are in the middle of Text 2, preceded by the title Sentence 1 and Sentences 2–10 about the Earth's predicament, and followed by Sentences 13–19 about the action to be taken to replenish our planet (in Text 2, the noun "Earth" is capitalised; therefore, for the sake of consistency, it is also capitalised in this Sect. 4.2). Of the definitions supplied by the Oxford English Dictionary and by the ecology and environment texts examined in Sect. 4.1, the most pertinent to the use of the marker word "environment" made in Text 2 are sense 2.d in the Oxford English Dictionary ("The natural world or physical surroundings in general, either as a whole or within a particular geographical area, esp. as affected by human activity") and the first half of the definition in Routledge's *The Environment Dictionary* (Kemp 2002, 127).

As shown in Sect. 4.1, the Oxford English Dictionary sense 2.d and its words "esp. as affected by human activity" acknowledges that we humans influence the environment and potentially damage it: consequently, it is beneficial and in line with the ecosophy of this research and its norm no. 5: Environmental Limits. The first half of *The Environment Dictionary* definition is beneficial and aligns with norms no. 1: Valuing Living and no. 2: Wellbeing, as it maintains that seemingly diverse environments are "all intimately linked and in combination constituting the whole earth/atmosphere system". On the other hand, the second half of *The Environment Dictionary* definition runs counter to the ecosophy and to four of its norms: no. 1: Valuing Living, no. 3: Now and the Future, no. 4: Care and no. 5: Environmental Limits. This part of the definition is hence destructive, since it adopts the technique of erasure (Stibbe 2021, 223; see Sect. 4.1) to background our human responsibility for dominating and destroying the environment, rather than adhering to the contrary technique of salience (Stibbe 2021, 227; see Sect. 4.1) to foreground our responsibility. By contrast, Text 2 deploys a large number of the stylistic mechanisms presented in the theoretical Sect. 2.2 to underline our human accountability for devastating nature.

The first of these stylistic mechanisms is foregrounding, more precisely semantic deviation (Short 1996, 43; see Sect. 2.2.1). Planet Earth is called "Mother Earth" (Sentences 2 and 10b) and anaphorically referred to by the feminine personal pronouns "she" and "her" (Sentences 3–5, 10a and 18). The planet is thus personified as a woman, to be more precise as a mother, a figurative use which is conventionalised and lexicalised (Goatly and LLE Project 2002–2005). In fact, "mother earth" is recorded in the Oxford English Dictionary as an independent entry, with the meaning of "The earth considered as the mother of its inhabitants and products. Also more generally: the ground". Furthermore, the many senses of the entry "mother" encompass sense 4.b, "The earth regarded as the source, nurturer, or sustainer of humanity". What is actually semantically deviant in Text 2 is the fact that Mother Earth is not described in accordance with the Oxford English Dictionary definitions, i.e. as protecting, nurturing and sustaining us humans. On the contrary, she is ascribed unexpected, negatively value-laden qualities, like "weary, disheveled and drained" (Sentence 3), and "cranky, frustrated and tired of humans" (Sentence 4); these qualities are set down in two three-part lists giving them additional prominence.

It is only in the concluding Sentences 18 and 19 that the Earth gradually becomes an environment and a planet again. The feminine personal pronoun "She" is employed in Sentence 18, but in Sentence 19 "Mother Earth" is merely referred to as "Earth" and, more importantly, as a "place". The unexpected, negatively value-laden adjectives in Sentences 3 and 4 are replaced by the more predictable and more value-neutral coordinated adjectives "old and complex" in Sentence 18, and by the positively value-laden adjective "fine" and adjective phrase "worth fighting for" in Sentence 19. The adjective phrase is the last in Text 2 and, as such, summarises the text and its entire message: strive determinedly for the planet and the environment. The stylistic strategy of foregrounding is here connected with the stylistic practice of point of view. The positive and negative evaluative adjectives and adjective phrase studied above convey the addresser's ideological point of view (Simpson 1993, 11–19, 27–39, 96–101; summarised in Neary 2014, 177–178; see Sect. 2.2.2) on the environment and her value-system by means of her unambiguous linguistic choices and selections (see Sect. 3.1 for a discussion of and references about the personification "Mother Nature" and Lovelock's [2009] Gaia hypothesis).

Following Sentences 1–5 devoted to the Earth, Sentences 6–14 are dedicated to us humans and to our interaction with the environment. This change in focus is relayed by another example of foregrounding: the repetition (Leech and Short 2007, 39–41; Mukařovský 1977; see Sect. 2.2.1) of the personal pronoun "we", which occurs eleven total times in the whole Text 2 (once in the title Sentence 1 and ten times in Sentences 6–14). In Sentence 6, which opens the human-centred sequence, we are depicted through the two coordinated adjectives "needy and greedy". Rhyming with each other, they constitute a case of phonological deviation (Short 1996, 54–55; see Sect. 2.2.1), which makes them salient to the addressee of Text 2, together with our unpleasant characterisation. Again, foregrounding is related to and reinforced by point of view. The personal pronoun "we" is a proximal deictic term utilised inclusively: in other words, it communicates that the angle and perspective the addresser writes Text 2 from comprise the addresser and all her addressees, viz. all of us humans (Simpson 2010, 295; see Sect. 2.2.2). Moreover, the adjectives "needy and greedy" express unequivocally the addresser's ideological point of view on us humans and our activities, which is strengthened by the use of further evaluative words and phrases in Text 2—see, for instance, "without remorse or restitution" (Sentence 7), "alarming" (Sentence 8) and "fatal" (Sentence 11).

Text 2 features two additional cases of repetition realising foregrounding: the modal auxiliary verb "must" is present six times (including once in the prominent title Sentence 1), and "need" is deployed as a modal lexical verb four times. Apart from the one in the title sentence, nine overall occurrences of "must" and "need" can be found in the eight Sentences 10b-17; more precisely, each sentence contains one of these two verbs, and Sentence 10b (the opening one in the sequence) contains both. "Must" and "need" introduce the lexical verbs signalling the actions we humans are obliged and required to pursue so that the environment can recover, for example "we must learn to reconnect with nature and consider the environment in everything we do" (Sentence 12). The even distribution of the two verbs in Text 2 furnishes it with a regular pattern, namely a list setting down clear and specific directives for us humans to comply with in order "to make the world resilient to future shocks" (Sentence 13) (for the use of another list in a beneficial ecological text, see Sect. 3.2 on Text 1). All the clauses comprising the two verbs and the directives they give are positive declarative clauses; this makes the directives practical and proactive, enabling us to control the environmental crisis by taking the initiative and by making positive change happen. The various directives in the list will be presented in more detail and assessed against the ecosophy in the evaluative Sect. 4.2.2.

Foregrounding, here repetition, is once again associated with and reinforced by point of view, here psychological point of view indicated by modality (Simpson 1993, 43–47; summarised in Neary 2014, 181–182, and in Sect. 2.2.2). The two verbs "must" and "need" are operators of deontic modality; that is, they suggest what the addresser of Text 2 considers to be our duties and commitments to the environment.

Text 2 also evokes its addresser's psychological point of view by means of other modal systems. The eight Sentences 10b-17, containing the nine instances of deontic modality, are preceded by Sentences 2-10a and followed by Sentences 18–19, showing six total occurrences of epistemic modality. This modal system conveys whether the addresser of Text 2 is assured of the truthfulness of the propositions she writes, viz. what she knows and believes about the environment. The six modal operators relaying epistemic modality in Sentences 2-10a are the following:

[2] estimated; [5] can; [6] always; [9] can; [10a] may; [18] may

Sentence 19 comprises the modal adverb "undoubtedly" ("But undoubtedly, Earth is a fine place and worth fighting for"): this is an operator of perception modality, which is a sub-system of epistemic modality based on visual powers of perception and related abilities. The perception modal operator therefore strengthens the stylistic role of the six epistemic modal operations, also owing to the fact that it signals a high degree of factuality, it has an outstanding position in the concluding sentence of Text 2, and it co-occurs with the linguistic indicators of ideological point of view "fine" and "worth fighting for" (see above).

As shown in Sect. 3.2.1, two of the six epistemic modal operators above can also be investigated in line with the Hallidayan standpoint on modality (Halliday 2014, 176–183). The first is the epistemic modal adverb "always" ("We've always been needy and greedy" in Sentence 6), which communicates a degree from the scale of usuality and oftenness "sometimes/usually/always"; accordingly, it is encompassed into the epistemic modal system. The second epistemic modal operator which can be scrutinised in agreement with Hallidayan modality is the modal auxiliary verb "can" ("Who can blame her?" in Sentence 5). Sentence 5 including the auxiliary verb is an interrogative clause; as the only one in Text 2, it realises a case of foregrounded internal deviation. The clause explicitly asks for the addressee's view on the Earth's unfavourable emotions about us humans; it can indeed be reworded as "Do you think somebody blames her [for feeling cranky, frustrated and tired of humans]?", which elicits the answer "No, I don't". In traditional terms, as mentioned in Sect. 3.2.1, the clause has the communicative function of a rhetorical question, namely a persuasive or emotive device making a direct appeal to the addressee, and explicitly involving them in the text and in its meaning-making processes (Wales 2011, 370–371). In Text 2, the addresser appeals to the addressee to blame us humans for our misbehaviour towards the environment.

As anticipated above, a clear pattern emerges from Text 2. Sentences 2-10a and 18–19 consider the Earth's predicament and are characterised by epistemic modality; they frame Sentences 10b-17, treating a list of activities to restore the environment and improve its resilience, typified by deontic modality. However, there are two exceptions to this topic and modality pattern, which constitute two further foregrounded cases of internal deviation. In Sentence 16 ("corporate interests that want to profit from it [sic] [forests]"), the modal lexical verb "want" is the only operator of boulomaic modality in Text 2, expressing wishes and desires, here those of the business corporations eager to exploit and ruin the forest environment. Nevertheless, stylistically, the only hint at corporate boulomaic modal wishes and desires is fiercely opposed by the nine deontic modal instances of the actions we humans must and need to carry out with a view to protecting not only forests, but the entire natural world.

The second foregrounded exception to the topic and modality configuration just outlined is the modal auxiliary verb "would" in Sentence 11 ("We need to stop treating the environment as a commodity and as an infinite resource because doing so would be fatal"). Firstly, it is an operator of epistemic modality occurring in Sentences 10b-17 characterised by deontic modality. Secondly, and more interestingly, it triggers counterfactuality (Fauconnier and Turner 2002; Riddle Harding 2007; see Sect. 2.2.2). "Doing so", i.e. indefinite continuation of "treating the environment as a commodity and as an infinite resource", is a future action belonging to a discourse world presented as disnarrated (Lambrou 2019; Prince 1988); that is to say, it is supposed or assumed to take place, but not necessarily due or expected to become real or true. In Text 2, the counterfactual action is assessed via the negatively value-laden adjective "fatal", a linguistic indicator of the addresser's ideological point of view (see above). Indeed, the major role of counterfactuality is to position the addressee by influencing their appraisal of given elements of a text; here, it signals that it is a matter of life and death to urgently adopt the list of measures suggested in the deontic modal Sentences 10b-17.

The deontic modal operators "must" and "need" evoke high degrees of modality and convey the addresser's strong commitment to our human duties and obligations to the environment; as a result, they provide Text 2 with positive modal shading (for modal shading, see Simpson 1993, 50–69; summarised in Neary 2014, 183–187, and in Sect. 2.2.2). Nearly all the epistemic modal operators in Text 2 relay medium degrees of modality and addresser commitment to the factuality of the Earth's predicament; in addition, the epistemic modal adverb "always" (Sentence 6) and the perception modal adverb "undoubtedly" (Sentence 19) communicate high degrees of modality and addresser commitment. Consequently, Text 2 is not supplied with negative modal shading, which is instead expressed by low degrees of modality. These modality patterns are reinforced by four sentences (Sentences 3, 4, 7 and 8) realising unmodalised categorical assertions, hence displaying neutral modal shading. The four sentences can be found in the sequence of Text 2 devoted to the Earth's predicament. As previously shown, Sentences 3 and 4 contain the two three-part lists of negative evaluative adjectives and adjective phrases portraying the planet ("weary, disheveled and drained"; "cranky, frustrated and tired of humans"). Conversely, Sentences 7 and 8 furnish the explanation for the Earth being assigned such unpleasant features: we humans are degrading the environment and exhausting its natural materials. None of these assertions is psychologically judged by the addresser; accordingly, they signal the highest possible level of addresser commitment to them and their veracity: it is a fact that the planet is tired of us humans, because we are draining her. In sum, a psychologically assertive addresser emerges from these configurations of modality and modal shading: an activist firmly dedicated to solving the environment's problems, to identifying their causes and to proposing practical solutions to them.

The three-part lists of adjectives and adjective phrases representing the Earth, as shown above, are two of the foregrounding and point of view stylistic techniques employed in Text 2 to conceptualise the Earth as a woman and as a mother. Conceptual metaphor (Lakoff and Johnson

2003; see Sect. 2.2.3) plays a significant part in this personification. Aligning with the three-dimensional model of metaphor developed by Steen (2016; see Sect. 2.2.3), the conceptualisation of the Earth as a woman and as a mother can be explored as follows:

1. At the level of expression, or metaphor in language, this personification is prompted by the noun phrase "Mother Earth" (Sentences 2 and 10b) and by the anaphoric use of the feminine personal pronouns "she" and "her" (Sentences 3–5, 10a and 18). Furthermore, a number of sentences concur in this personification:
   1.a. Sentences 3 and 4 encompass the two three-part lists;
   1.b. The expressions "what she's been through" (Sentence 4), "Who can blame her?" (Sentence 5), "we need to give Mother Earth a break—we must offer her a lifeline" (Sentence 10b) all presuppose a sentient participant in the processes the expressions indicate; in particular, "we must offer her a lifeline" presupposes the existence of an animal, if not human, body able to catch and cling on to a line or rope.
2. At the level of conceptualisation, or metaphor in thought, the noun phrase "Mother Earth" features both the source MOTHER and the target THE EARTH. This brings about the conceptual metaphor THE EARTH IS A MOTHER, along with the metaphorical entailment (Stibbe 2021, 68) that the Earth is a woman, therefore a sentient being with a human body and human physical and mental sensations.
3. At the level of communication, or metaphor in interaction, as previously mentioned, THE EARTH IS A MOTHER is a conventionalised and lexicalised metaphor recorded in the Oxford English Dictionary. As a result, it is not a creative metaphor with a deliberate figurative function.

Although conventionalised and lexicalised, the metaphor is not reported in the three databases Master Metaphor List (Lakoff and Cognitive Linguistics Group 1991), Metalude (Goatly and LLE Project 2002–2005) and MetaNet Metaphor Wiki (David et al. 2016, 2018). However, its source MOTHER is a role name in the MetaNet frame Nurturant Parent Family (for frames, see Tannen 1993. In accordance with common typographical conventions, the initials of frames are Capitalised, and metaphors, targets and sources are in SMALL CAPITALS). The frame also includes the role name "Parenting Style", typified by the definition/comments "Focus on empathy, need-based fairness", and the role name "Traits in Children", characterised by the definition/comments "Focus on empathy, tolerance, social responsibility". These linguistic data confirm the findings of the foregrounding and semantic deviation analysis above. What is creative and deliberate in Text 2 and in the THE EARTH IS A MOTHER conceptual metaphor is the fact that the conceptual metaphor is not in line with the "Parenting Style" role name and its definition. In other words, the target THE EARTH is not described in the prototypical terms of the source MOTHER ("Focus on empathy, need-based fairness"), but in less prototypical and more unusual and unfamiliar terms (for the concept of defamiliarisation, see Emmott and Alexander 2016, 291; Gavins 2014; Šklovsky 2012; and Sect. 2.2.1). Consequently, the conventionalised and lexicalised THE EARTH IS A MOTHER metaphor must be rewritten as the creative and deliberate metaphor THE EARTH IS A SHREWISH MOTHER.

If THE EARTH IS A MOTHER, it follows that WE HUMANS ARE THE EARTH'S CHILDREN. This conceptual metaphor can be examined by taking into account the "Traits in Children" role name and its definition/comments "Focus on empathy, tolerance, social responsibility". Nevertheless, Text 2 unmistakably suggests that our human behaviour does not observe this role name or this definition, and that we are anything but empathetic, tolerant or socially responsible with regard to the environment (see, for example, Sentences 6–9 and 11 about our human exploitation of the planet). The WE HUMANS ARE THE EARTH'S CHILDREN conceptual metaphor must hence be rephrased as WE HUMANS ARE THE EARTH'S DEGENERATE CHILDREN. This rephrased conceptual metaphor accounts for the THE EARTH IS A SHREWISH MOTHER conceptual metaphor. The reason the mother is shrewish is that her children are degenerate; accordingly, she is vindicated for her being unusually "cranky, frustrated and tired of humans".

On the one hand, Text 2 conceptualises us humans as the Earth's degenerate children; on the other hand, the text depicts us as having the responsibility to reclaim the environment through deontic modality and the list of resilient courses of action in Sentences 10b-17. However, Sentence 10b, the one opening the deontic modal sequence ("we need to give Mother Earth a break—we must offer her a lifeline"), is problematic. Underpinning the finite clause "we must offer her a lifeline" are two more conceptual metaphors, which can be studied as follows:

1. At the level of expression, or metaphor in language, the noun "lifeline" denotes a line or rope utilised in life-saving. The fact that "we [humans] must offer her [the Earth] a lifeline" entails that the personified planet is in need of a lifeline, which in turn entails that she is in danger of drowning in the sea. That the lexical item "lifeline" can also be deployed metaphorically is confirmed by its occurrence in two Metalude root analogies, ACTIVITY IS SWIMMING AND FAILURE IS SINKING. The lexical item has the literal meaning "rope thrown to a drowning person / a person drowning in the sea", and the metaphorical meanings "means of continuing with life or an activity" and "way of getting help to continue your life".
2. At the level of conceptualisation, or metaphor in thought, "her" constitutes the target THE EARTH; the personified planet's need of a lifeline and the danger she is in activate the source A DROWNING WOMAN. This results in the first conceptual metaphor THE EARTH IS A DROWNING WOMAN. Moreover, "we" realises the target WE HUMANS in another conceptual metaphor; our action of offering the planet a lifeline and saving her life by means of it triggers the source THE EARTH'S RESCUERS. This leads to the second conceptual metaphor WE HUMANS ARE THE EARTH'S RESCUERS.
3. At the level of communication, or metaphor in interaction, both THE EARTH IS A DROWNING WOMAN and WE HUMANS ARE THE EARTH'S RESCUERS are creative metaphors with a deliberate figurative function. Neither is, in fact, set down in the three metaphor databases cited above.

But why is Sentence 10b questionable, as well as these two conceptual metaphors it is founded on? THE EARTH IS A DROWNING WOMAN is apparently similar to THE EARTH IS A SHREWISH MOTHER, given that underlying both conceptual metaphors is THE EARTH IS A

PERSON/WOMAN. As summarised in Stibbe (2021, 66–67), this is an ambivalent metaphor (see Sect. 2.1.3), so much so that some authors comment on its beneficial ecocentric characteristics, some others on its destructive anthropocentric traits. On the one hand, the conceptual metaphor is regarded as placing the organism of the Earth as the focus of human and more-than-human life and existence. On the other hand, the conceptual metaphor and its gendering of the planet as a woman are reckoned to remind the addressee of the male patriarchal domination of women and to make it analogous to our human domination of the Earth (Berman 1994). In the context of Text 2, THE EARTH IS A SHREWISH MOTHER is a beneficial and empowering metaphor, as it personifies the planet as a self-aware woman capable of strong unmixed feelings. By contrast, the THE EARTH IS A DROWNING WOMAN conceptual metaphor effectively alludes to the environmental crisis and to its being acute, immediate and full-scale; it is still ambivalent, since it also portrays the planet as a powerless woman oppressed by us humans and risking death for this.

With regard to WE HUMANS ARE THE EARTH'S RESCUERS, as evoked by Text 2 and its deontic modal Sentences 10b-17, we are the only species who can and must replenish the environment and make it more resilient via the list of steps catalogued in the sentences. Nevertheless, we humans having the power "to give Mother Earth a break", rescuing the planet and throwing a lifeline to a powerless drowning woman, thereby saving her life, are represented as heroic and capable of exerting great force and producing great effects; that is to say, we are conceptualised in extremely agreeable terms we are unworthy of. In addition, Text 2 unambiguously conveys our human responsibilities to the environment in Sentences 6–9 through several stylistic and foregrounding mechanisms (see above). The finite clause "we must offer her a lifeline" and the resulting conceptual metaphor WE HUMANS ARE THE EARTH'S RESCUERS, in contrast, are based on the backgrounding device of erasure. What the clause and the conceptual metaphor try to erase is our rapacious and remorseless personality and our excessive capitalising on natural materials. More importantly, what they also try to erase is the fact that we humans are a part of the Earth and its environment: if she metaphorically drowns in the sea, we all go down with her.

4.2.2 Summary and Evaluation

Text 2, taken from Greenpeace's website and comprising the marker word "environment", and the stylistic strategies of foregrounding, point of view and metaphor figuring in the text will be summarised and evaluated in this section, following the ecostylistic investigation of the text and its stylistic practices in the previous section. To be more precise, the value-system of Text 2 emerging from this investigation is assessed here against the ecosophy of this research and its norms; it is therefore demonstrated whether the text and the techniques employed by the addresser are actually beneficial.

As shown in Sect. 4.2.1, underpinning Text 2 and its notion of environment are the beneficial definition of this marker word in the Oxford English Dictionary (OED Online 2022), mostly sense 2.d., and the first half of the definition in Routledge's *The Environment Dictionary* (Kemp 2002, 127); in turn, underpinning the two definitions is the ecosophy and its norms no. 1: Valuing Living, no. 2: Wellbeing and no. 5: Environmental Limits. In other words, Text 2 relays the ecocentric message that "Mother Earth" and the lives of all the human, animal and vegetable organisms she hosts are worthy in themselves, have the right to be healthy and must be conserved by not using natural materials to excess. This message is mainly communicated by the unusual description of the planet as a shrewish mother and by the detailed list of concrete measures to restore the environment: as a result, both the description and the list are beneficial stylistic mechanisms which should be propagated among the public at large. They are shaped by a large number of stylistic devices, which often co-occur in Text 2 at clause level, phrase level and even word level: the stylistic technique of foregrounding is frequently linked to the stylistic mechanisms of point of view and metaphor to underscore parts of the text and to make them more effective.

For instance, foregrounding consists of the semantic deviation "Mother Earth" (Sentences 2–5, 10 and 18) and of the two three-part lists of semantically deviant adjectives and adjective phrases "weary, disheveled and drained" (Sentence 3), and "cranky, frustrated and tired of humans" (Sentence 4); it is utilised to personify the planet as a woman and a mother figure. This personification is founded on the conventionalised and lexicalised metaphor THE EARTH IS A MOTHER; the metaphor is recorded in the Oxford English Dictionary entries "mother earth" and "mother", which define the Earth as a mother figure nourishing, supporting and cherishing us humans. The logical consequence is that WE HUMANS ARE THE EARTH'S CHILDREN. Text 2, however, depicts Mother Earth in an unusual and negatively value-laden way. In fact, the planet is not portrayed as a mother figure aligning with the Oxford English Dictionary definitions. Conversely, the planet is portrayed as a surprisingly bad-tempered woman understandably weary with us humans and with our unecological behaviour and actions: this novel mother figure is based on the creative and deliberate metaphor THE EARTH IS A SHREWISH MOTHER. Due to our behaviour and actions, the logical consequence must be modified accordingly as WE HUMANS ARE THE EARTH'S DEGENERATE CHILDREN. Because of their substantial defamiliarising impact on the addressee, these foregrounding and conceptual strategies imprint highly abiding and potent conceptual metaphors of the planet and us humans on the addressee's mind.

These two creative and deliberate metaphors are beneficial stylistic practices in accordance with the ecosophy of this research. More precisely, the Earth is represented as a mother angered at her degenerate human children, given that all our actions and activities run counter to the ecosophy and its norms. The norm the two conceptual metaphors are founded on and we humans are primarily diverging from is no. 5: Environmental Limits: the amazing amount of all sorts of goods we consume and squander is harming the ability of the planet and of the environment to support humans, animals and plants and our common flourishing in the near and distant future alike. Underlying the conceptualisation of the Earth as a bad-tempered mother and of us humans as her debased children is also norm no. 3: Now and the Future: with a view to us rebuilding our impaired parent-children relationship, we must realise that we have to face ecological challenges, deal with the environmental crisis and draw up a plan of action without delay.

This is because, as Text 2 puts it, "Earth is a fine place and worth fighting for" (Sentence 19). Here, the planet is attributed the favourable evaluative adjective "fine" and adjective phrase "worth fighting for". Together with the unfavourable value-laden adjectives and adjective phrases in the two three-part lists mentioned above, and with several other evaluative words and phrases in Text 2, the positively and negatively value-laden words and phrases in the text express the addresser's ideological point of view on the Earth, us humans and the environment. Their role is to position the addressee and to make them share the addresser's worldview and ecosophy, principally its norms no. 3: Now and the Future and no. 5: Environmental Limits.

A number of stylistic devices highlight the facts that we humans must tackle the environmental crisis now and that we are overstepping the planet's environmental limits; such devices are therefore beneficial when compared with the ecosophy. Among them are the foregrounded repetition of the proximal deictic personal pronoun "we", deployed inclusively eleven times (Sentences 1 and 6–14), and the equally foregrounded phonological deviation in the two coordinated negatively value-laden adjectives "needy and greedy" (Sentence 6). These strategies signal that all of us humans are the agents to blame for creating the environmental crisis, which is why we are characterised and appraised in a negative way; nevertheless, we are all also the agents responsible for solving this crisis. This blame and this responsibility are indicated by two further foregrounded repetitions, that of the modal auxiliary verb "must" (six times) and that of the modal lexical verb "need" (four times) in Sentences 1 and 10b-17. The two verbs precede the lexical verbs suggesting the directives and courses of action we humans are requested to take to reclaim the environment, resulting in a thorough and practical list.

Of the steps contained in the list, the following are those which most openly agree with the ecosophy and its norms:

1. "We need to stop treating the environment as a commodity and as an infinite resource because doing so would be fatal" (Sentence 11); and "Efforts to protect at least 30% of the world's oceans by 2030 must be supported to save marine biodiversity and to help save the climate" (Sentence 15): these two steps are in line with norm no. 1: Valuing Living, which claims that the entire ecosphere has intrinsic value, with all its human and more-than-human life, including marine life and biodiversity; consequently, the environment must not be treated "as a commodity and as an infinite resource", but must be protected and

saved (see also the points below) (for an ecostylistic examination of the term "resource", see Sect. 6.2).

2. "we must learn to reconnect with nature and consider the environment in everything we do" (Sentence 12): this step is in line with norm no. 2: Wellbeing, noting that we humans must get rid of our "unessential baggage" and "useless diversions" (Odum and Odum 2001, 3, 4), and must value unsophisticated lifestyles and pleasures, such as "reconnect[ing] with nature". This step also follows norm no. 4: Care, observing that we humans must always bear the environment in mind and deal with it with empathy, regret and gratitude.
3. "governments need to shift their priorities and invest on [sic] human and planetary health to make the world resilient to future shocks" (Sentence 13): this step is in line with norm no. 7: Adaptation, which points out that we humans must tailor ourselves and our ways of life to new ecological circumstances, and must enhance our resilience to the environmental changes to come. This step also follows norm no. 1: Valuing Living, which states that our human health must be on the same level as planetary health.
4. "We need to decarbonise economies and shift to renewable energy systems to mitigate global warming" (Sentence 14): this step is in line with norm no. 3: Now and the Future, arguing that we humans must take at once comprehensive and hardline measures aimed at replenishing the environment.
5. "Forests that have stood the test of time and teeming with life must be preserved and protected from corporate interests that want to profit from it [sic]" (Sentence 16): this step is in line with norms no. 1: Valuing Living, for the reasons shown above, and no. 6: Social Justice. The latter norm contends that deforestation is one of the threats constituting the phenomenon of slow violence (Nixon 2011), which must also be handled by altering the present economic and social situation and by making new and fairer policies (see also point 6 below).
6. "Indigenous Peoples, with their ancient wisdom of the natural world[,] must be empowered and recognised" (Sentence 17): this step is in line with norm no. 6: Social Justice, maintaining that two of the plans to confront slow violence against poor peoples from the Global

South are to understand and regard Indigenous Peoples highly, and to increase the role and power of their local identities, qualities and practices (for the issue of Indigenous nations and peoples in beneficial ecological texts, see also Text 1 and Text 4 and Sects. 3.2 and 6.2).

The beneficial stylistic technique of the list of environmental courses of action is based not only on the foregrounding mechanism of repetition, but also on the psychological point of view device of modality; the two verbs "must" and "need", repeated in Text 2, also evoke deontic modality. To be more precise, the operators of this modal system supply the text with positive modal shading: they communicate high values of modality expressing the addresser's firm commitment to what we humans must and need to do so that the environment can recover. The addresser's psychological point of view is also conveyed via epistemic modality (six cases in Sentences 2-10a and 18), boulomaic modality (the lexical verb "want" in Sentence 16; see point 2 in the numbered list below) and perception modality (the adverb "undoubtedly" in Sentence 19). With regard to the epistemic and perception modal systems, their modal operators do not furnish Text 2 with negative modal shading: on the contrary, their medium to high values of modality signal the addresser's unflinching commitment to the truthfulness of the Earth's predicament. The factuality indicated by positive modal shading and by the lack of negative modal shading is confirmed by neutral modal shading. This is present in four unmodalised categorical assertions (Sentences 3, 4, 7 and 8), where the addresser notes that the planet is in a difficult situation and that all its causes are human-induced. These patterns of modality and psychological point of view characterise the addresser of Text 2 as a person resolute in pursuing her environmental and social ends, and confident of the judgements she makes in the text. These assertive practices inspire trust in her addressee and must accordingly be deployed by ecological speakers and writers to circulate the ecosophy of this research and its norms.

Three modal operators and the clauses they occur in are foregrounded through internal deviation; hence, the propositions they psychologically judge and their stylistic strategies become particularly conspicuous, and so do their connections with the ecosophy and its norms:

1. The epistemic modal auxiliary verb "can", in "Who can blame her?" (Sentence 5), is employed in the only interrogative clause in Text 2. The clause is a rhetorical question encouraging the addressee to concur with the addresser that the Earth cannot be blamed for being a metaphorical shrewish mother, as we humans are her figurative degenerate children, viz. the only species accountable for the environmental crisis. This aligns with norm no. 5: Environmental Limits.
2. The noun phrase "corporate interests that want to profit from it [sic] [forests]" (Sentence 16) features the modal lexical verb "want", which is the only boulomaic modal operator in Text 2. The verb is utilised to relay what business corporations want to do to the forest environment; however, the boulomaic corporate desires and wishes are stylistically outnumbered and transcended by the deontic list of steps we humans must and need to perform in order to restore the environment. This is in accordance with norms no. 1: Valuing Living and no. 6: Social Justice (see also point 5 in the numbered list above).
3. The epistemic modal auxiliary verb "would", in "We need to stop treating the environment as a commodity and as an infinite resource because doing so would be fatal" (Sentence 11), prompts the only case of counterfactuality and disnarration in Text 2. In other words, the "fatal" action of "doing so" is simply presumed at this stage and will come true if we humans do not obey the ecosophy and live in line with it, especially with its norm no. 1: Valuing Living (see also point 1 in the numbered list above).

So far, all the stylistic techniques detected in Text 2 and evaluated against the ecosophy have proven to be beneficial and to follow its norms. Nevertheless, Sentence 10b ("we need to give Mother Earth a break—we must offer her a lifeline") includes two non-beneficial stylistic mechanisms elicited by the noun "lifeline": the creative and deliberate metaphors THE EARTH IS A DROWNING WOMAN and WE HUMANS ARE THE EARTH'S RESCUERS. As shown in Sect. 4.2.1, the former conceptual metaphor is ambivalent. On the one hand, it beneficially represents the environmental crisis as full-blown and to be solved immediately, thus aligning with norm no. 3: Now and the Future. On the other hand, the conceptual metaphor destructively conceptualises the planet as an entity brought under human domination and control, thereby running counter to norms no. 1: Valuing Living and no. 4: Care: the only lives to be valued are our human lives, and no care is due to nonhuman organisms. As far as the WE HUMANS ARE THE EARTH'S RESCUERS conceptual metaphor is concerned, it is fully destructive. It describes us as heroes saving a woman's life, but it erases at least two facts:

1. It is our being exceedingly "needy and greedy" for the planet's natural materials that is to blame for provoking the environmental crisis, namely our diverging from norms no. 1: Valuing Living and no. 5: Environmental Limits;
2. We humans live on the Earth along with other animals and plants; as a result, we are not independent, separate or even alienated from them, but we share their ecological predicament; erasing this fact goes against norms no. 1: Valuing Living and no. 4: Care.

Consequently, destructive Sentence 10b must be reworded in a beneficial way. The first finite clause "we need to give Mother Earth a break" depicts us humans in appealing terms, i.e. as empowered entities who can choose whether to stop putting pressure on the planet and treating her in a harsh way. These appealing terms can be replaced with the less appealing terms of one of the two beneficial metaphors underpinning Text 2 and emerging from this ecostylistic scrutiny: "we need to stop being the Earth's degenerate children". The second finite clause "we must offer her a lifeline" comprises the noun "lifeline", which activates the ambivalent metaphor THE EARTH IS A DROWNING WOMAN and the destructive metaphor WE HUMANS ARE THE EARTH'S RESCUERS. The clause is not particularly easy to rewrite without this keyword; hence, it can be left out of a beneficial version of the text: "Though she may be resilient, we need to stop being the Earth's degenerate children". As revealed by analysing Text 2 from an ecostylistic perspective, and by comparing its message with the ecosophy of this research, ecological and social change may be brought about by ground-breaking stylistic strategies portraying us humans and our interplay with the planet, and by a list of concrete measures to reclaim the environment, as advised by Greenpeace's environmental website.

References

Bailey, J., ed. 2004. *The Facts on File Dictionary of Ecology and the Environment*. New York: Market House Books.

Berman, T. 1994. The rape of Mother Nature? Women in the language of environmental discourse. *The Trumpeter* 11 (4): 173–178.

Collin, P.H. 2011 [1985]. *Dictionary of Environment & Ecology*, 5th edition. London: Bloomsbury [electronic edition].

David, O., G. Lakoff et al. 2016–2018. *MetaNet*. Berkeley: University of California. https://metaphor.icsi.berkeley.edu/pub/en. Last accessed April 2022.

Emmott, C., and M. Alexander. 2016. Defamiliarization and foregrounding: Representing experiences of change of state and perception in neurological illness autobiographies. In *The Bloomsbury Companion to Stylistics*, ed. V. Sotirova, 289–307. London: Bloomsbury.

Fauconnier, G., and M. Turner. 2002. *The Way We Think: Conceptual Blending and the Mind's Hidden Complexities*. New York: Basic Books.

Gavins, J. 2014. Defamiliarisation. In *The Cambridge Handbook of Stylistics*, ed. P. Stockwell and S. Whiteley, 196–211. Cambridge: Cambridge University Press.

Goatly, A., and LLE Project. 2002–2005. *Metalude*. Hong Kong: Lingnan University. http://www.ln.edu.hk/lle/cwd/project01/web/home.html. Last accessed April 2022.

Halliday, M.A.K. 2014 [1985]. *Halliday's Introduction to Functional Grammar*, revised by C.M.I.M. Matthiessen, 4th edition. London and New York: Routledge.

Kemp, D.D. 2002 [1998]. *The Environment Dictionary*, 2nd edition. London and New York: Routledge [electronic edition].

Lakoff, G., and Cognitive Linguistics Group. 1991. *Master Metaphor List*. Berkeley: University of California. http://araw.mede.uic.edu/~alansz/metaphor/METAPHORLIST.pdf. Last accessed April 2022.

Lakoff, G., and M. Johnson. 2003 [1980]. *Metaphors We Live By: With a New Afterword*, 2nd edition. Chicago: University of Chicago Press.

Lambrou, M. 2019. *Disnarration and the Unmentioned in Fact and Fiction.* Basingstoke: Palgrave Macmillan.

Larson, B. 2011. *Metaphors for Environmental Sustainability: Redefining our Relationship with Nature.* New Haven, CT: Yale University Press.

Leech, G., and M. Short. 2007 [1981]. *Style in Fiction: A Linguistic Introduction to English Fictional Prose*, 2nd edition. Harlow: Longman.

Lovelock, J. 2009. *The Vanishing Face of Gaia: A Final Warning.* New York: Basic Books.

Miller, G.T., and S.E. Spoolman. 2019 [2007]. *Essentials of Ecology*, 8th edition. Belmont, CA: Cengage.

Mukařovský, J. 1977. On poetic language. In *The Word and Verbal Art: Selected Essays*, ed. J. Burbank and P. Steiner, 1–64. New Haven, CT and London: Yale University Press.

Myerson, G., and Y. Rydin. 1996. *The Language of Environment: A New Rhetoric.* London and New York: Routledge.

Neary, C. 2014. Stylistics, point of view and modality. In *The Routledge Handbook of Stylistics*, ed. M. Burke, 175–190. London and New York: Routledge.

Nixon, R. 2011. *Slow Violence and the Environmentalism of the Poor.* Cambridge, MA: Harvard University Press.

Odum, H.T., and E.C. Odum. 2001. *A Prosperous Way Down: Principles and Policies.* Boulder, CO: University Press of Colorado.

Oxford English Dictionary (OED) Online. 2022. Oxford: Oxford University Press. https://www.oed.com/. Last accessed April 2022.

Prince, G. 1988. The disnarrated. *Style* 22: 1–8.

Riddle Harding, J. 2007. Evaluative stance and counterfactuals in language and literature. *Language and Literature* 16 (3): 263–280.

Short, M. 1996. *Exploring the Language of Poems, Plays and Prose.* Harlow: Longman.

Simpson, P. 1993. *Language, Ideology and Point of View.* London and New York: Routledge.

Simpson, P. 2010. Point of view. In *Language and Style: In Honour of Mick Short*, ed. D. McIntyre and B. Busse, 293–310. Basingstoke: Palgrave MacMillan.

Šklovsky, V. 2012 [1965]. Art as technique. In *Russian Formalist Criticism: Four Essays*, 2nd edition, eds. L.T. Lemon and M.J. Reis, 3–24. Lincoln and London: University of Nebraska Press (first published in Russian in 1917).

Smith, T.M., and R.L. Smith. 2015. *Elements of Ecology*, 9th ed. Harlow: Pearson.

Steen, G. 2016. Metaphor: Metaphor and style through genre, with illustrations from Carol Ann Duffy's *Rapture*. In *The Bloomsbury Companion to Stylistics*, ed. V. Sotirova, 308–324. London: Bloomsbury.
Stibbe, A. 2021 [2015]. *Ecolinguistics: Language, Ecology and the Stories We Live By*, 2nd edition. London and New York: Routledge.
Tannen, D., ed. 1993. *Framing in Discourse*. Oxford: Oxford University Press.
Wales, K. 2011 [1990]. *A Dictionary of Stylistics*, 3rd edition. London and New York: Routledge.

5

Ecosystem

5.1 Definition and Evaluation

The Oxford English Dictionary (OED Online 2022) supplies the following definition of the marker word "ecosystem", which is the third of the five marker words (Myerson and Rydin 1996, 6; see Sect. 1.1) explored in this book:

> A biological system composed of all the organisms found in a particular physical environment, interacting with it and with each other. Also in extended use: a complex system resembling this.
> (OED Online 2022)

In other words, an ecosystem is an organised and interconnected network of natural organisms perceived as a unity and forming a complex whole. As is the case with the definition of the noun "environment" studied in Sect. 4.1, that of "ecosystem" also entails equilibrium and a condition of balance among all the various biological and physical organisms and forces in the system; this condition must be preserved and protected with a view to also preserving and protecting the life and

D. F. Virdis, *Ecological Stylistics*,
https://doi.org/10.1007/978-3-031-10658-3_5

wellbeing of the entire system. On these grounds, like "environment", "ecosystem" as delineated above is a beneficial term (see Sect. 2.1.2) when assessed against the ecosophy of this research and its norms no. 1: Valuing Living and no. 2: Wellbeing (see Sect. 1.3).

A definition of the noun "ecosystem" appears in all the recently published or reprinted ecology and environment dictionaries and textbooks I consulted. Oxford's *A Dictionary of Ecology* (Allaby 2015, 136) features a long and many-sided one:

> ecosystem (ecological system) A term first used in 1930 by Arthur Roy Clapham (1904-90) and popularized by A.G. Tansley (in 1935) to describe a discrete unit that consists of living and non-living parts, interacting to form a stable system. Fundamental concepts include the flow of energy via food-chains and food-webs, and the cycling of nutrients biogeochemically (see BIOGEOCHEMICAL CYCLE). Ecosystem principles can be applied at all scales. Principles that apply to an ephemeral pond, for example, apply equally to a lake, an ocean, or the whole planet. In Russian and central European literature "biogeocoenosis" describes the same concept. (Allaby 2015, 136)

This definition describes several aspects of the ecosystem that merit discussion, a number of which are very similar to, if not the same as, the beneficial qualities of the noun "environment". An ecosystem is simultaneously depicted as a distinct whole and as a group of individual entities constituting a solid and ordered structure by means of their interconnectedness. On the other hand, the Russian and central European noun "biogeocoenosis" underscores the intimate association ("coenosis") between living entities ("bio-") and their physical environment ("geo-"). The noun "ecosystem" can be utilised on both a micro-scale and a macro-scale, and its major property of interconnectedness or intimate association among its components has relevance at all these diverse extents.

This property is conceptualised through the use of the conventionalised and lexicalised metaphors (Goatly and LLE Project 2002–2005) "food-chains", "food-webs" and "the cycling of nutrients". The noun "chain" in the compound "food-chain" figuratively denotes a connected series, sequence or set of entities; as a result, a "food-chain" is a series of organisms each linked to and dependent upon the next for food. In the compound "food-web", the noun "web" metaphorically indicates an intricate or interconnected structure or series of elements or, in a word, a network; a "food-web" is, hence, the system of interdependent food-chains in a community. The noun "cycling", in the noun phrase "the cycling of nutrients", designates the figurative action of moving in or passing through cycles; consequently, the noun phrase portrays the pathways of the elements through the ecosystem (from assimilation to decomposition) as if completing cycles. Considering this figurative evidence, the notion of interconnectedness is embraced in the conceptual metaphor (Lakoff and Johnson 2003; see Sect. 2.2.3) NATURE IS A WEB. This is a stylistic technique expressing the idea of a web having interconnected threads and of us humans being just one thread among countless others; it was accordingly recognised as a beneficial metaphor (Stibbe 2021, 67–68; see also Sect. 7.2. For sustainable (or beneficial) metaphors and unsustainable (or destructive) metaphors, see Larson 2011).

However, the beneficial noun "ecosystem" is also deployed in the destructive noun phrase (see Sect. 2.1.4) "ecosystem services"; among the texts referenced here, it is defined in *Essentials of Ecology* (Miller and Spoolman 2019, G4) and in the *Bloomsbury Dictionary of Environment & Ecology* (Collin 2011, 69). The most destructive version of this definition occurs in the former:

> **ecosystem services** Natural services or natural capital that support life on the earth and are essential to the quality of human life and the functioning of the world's economies. Examples are the chemical cycles, natural pest control, and natural purification of air and water. See *natural resources*. (Miller and Spoolman 2019, G4)

As shown in the scrutiny of the noun phrase "natural services" in Sect. 3.1, the noun "service" denotes a particular type of activity required by others and contributing to their welfare, or a facility supplying some public demand or providing maintenance and repair. In the definition of "ecosystem services", the so-called services furnished by the ecosystem are signalled by the two coordinated finite clauses "that support life on the earth" and "are essential [...] world's economies". The two clauses raise certain issues. Firstly, the fact that they are coordinated means that the information each of them indicates is of equal rank and importance; that is to say, the entire life on the planet is represented as being tantamount to the world's economic systems. Secondly, in the first clause, the noun phrase "life on the earth" incorporates all living beings (humans, animals and plants) collectively and without exception; in the second clause, this collective scope narrows, so much so that the beings benefitting from what the ecosystem can offer are us humans and our "human life" only.

As mentioned in Sect. 3.1, the definition of the noun phrase "natural services" diverges from at least two norms in the ecosophy of this research, i.e. no. 1: Valuing Living (appreciating all human and more-than-human life) and no. 4: Care (empathy with, regret at and gratitude to all animal and vegetable species). Like "natural services", the noun phrase "ecosystem services" and its definition do not comply with these two norms. The noun phrase conceptualises the ecosystem and the nonhuman organisms composing it as helpful materials to be exploited, not as being an end in themselves: it is therefore destructive. A beneficial rewriting of the definition would not employ the nouns "services" or "capital", and would replace them with such nouns as "processes", "actions" or "operations". Moreover, as established by the ecosophy and its norm no. 2: Wellbeing, it is not "the quality of human life" only that must be sustained and improved, but the welfare of all humans, animals and plants; as a result, the adjective "human" would be left out of a beneficial version of the definition. Another part which would be left out is the noun phrase "the functioning of the world's economies". This is because, in line with norm no. 2: Wellbeing, one of our ideological tenets must be physical and moral welfare, not material welfare or economic preoccupations. In addition, both norms no. 6: Social Justice and no. 7: Adaptation point out that our economic systems must be profoundly modified and new policies must be developed, for the present ones, hinted at in the definition, are collapsing (for a beneficial alternative, see Sect. 7.2.1 and the notion of "the living processes which ensure food and water for all").

5.2 "Ecosystem" in Forestry England's Words

5.2.1 Ecostylistic Analysis

The marker word "ecosystem" has now been delineated from the lexicological perspective of the Oxford English Dictionary (OED Online 2022) and from the naturalistic perspective of ecology and environmental studies. This section explores how this marker word is utilised in the environet (Myerson and Rydin 1996, 7; see Sect. 1.1), specifically in the environmental website and discourse of Forestry England (https://www.forestryengland.uk/) (see Sect. 1.2), and what beneficial ecostylistic devices are related to it. Text 3 is taken from the article "How beavers are returning to England's forests". The text consists of the article's opening section (Sentences 1–11) and of its second section (Sentences 13–23); the latter is headed by Sentence 12.

Text 3

[1] Walking along the water's edge, small tree stumps dot the area. [2] The wood has been chiselled into a pencil-like point, as if sharpened with a knife. [3] Webbed hind footprints in the mud are another sign that a forester did not fell these trees. [4] They were felled by one of nature's greatest engineers, the beaver.

[5] The Eurasian beaver (*Castor fiber*) is a large semi-aquatic mammal and one of the largest rodents in the world. [6] Their thick, waterproof fur and broad, leathery tail make them excellent swimmers. [7] Known for tree felling and dam building, beavers have huge orange teeth hardened with iron that they used [sic] to "coppice" waterside trees by gnawing on the stems.

[8] These charismatic creatures were once widespread throughout Britain. [9] However, beavers were heavily hunted for their fur and scent glands that produce a substance called castoreum. [10] By the beginning of the 16th century the beaver became extinct from the country. [11] Their absence has changed landscapes ever since.

[12] Ecosystem Engineers

[13] Beavers play an important role in complex wetland ecosystems, creating habitats for many other plant, insect and mammal species. [14]

Few other animals have the ability to modify and shape their surrounding environment like the beaver does. [15] Their damming creates complex pools and riffles, providing both deeper water and shallow, fast-flowing areas, important for a range of aquatic life. [16] The coppicing they do opens up the canopy, creating areas of vegetation great for breeding birds. [17] How they feed increases deadwood in the watercourse, which leads to an abundance of invertebrates. [18] For these reasons, beavers are often referred to as "ecosystem engineers" and known as a keystone species. [19] Where beavers return, they can help to restore an entire ecosystem.

[20] Their return can benefit people too. [21] Beaver dams increase water storage and slow the flow of water downstream, potentially reducing the impact of flooding in the surrounding area. [22] The dams are a real feat of engineering in themselves. [23] The largest are 2-metre-high woven structures of branches and vegetation packed with mud to hold back the water.

(Crown Copyright, courtesy Forestry Commission (no date of publication), licensed under the Open Government Licence. Available at https://www.forestryengland.uk/blog/how-beavers-are-returning-englands-forests. Last accessed April 2022)

[233 words]

The marker word "ecosystem" occurs four times in Text 3, in four different sentences: the heading "Ecosystem Engineers" (Sentence 12), "Beavers play an important role in complex wetland ecosystems [...]" (Sentence 13), "For these reasons, beavers are often referred to as 'ecosystem engineers' and known as a keystone species" (Sentence 18) and "Where beavers return, they can help to restore an entire ecosystem" (Sentence 19). These occurrences organise Text 3 into three sequences:

1. The first sequence (Sentences 1–11) corresponds to the article's opening section and provides background information about beavers;
2. The second sequence (Sentences 12–19) contains the four "ecosystem" occurrences; one of them is the first word in the heading Sentence 12, and another the last word in the concluding Sentence 19. The sequence explains the ecological role of beavers in the ecosystems they live in, and why they are often called "ecosystem engineers";

3. The third section (Sentences 20–23) describes the dams built by beavers and their benefits for the ecosystem.

Section 5.1 studied the definitions of the marker word "ecosystem" present in the Oxford English Dictionary and in the ecology and environment dictionaries and textbooks referenced in this book. The instances of this marker word in Text 3 are founded on the Oxford English Dictionary definition: "A biological system composed of all the organisms found in a particular physical environment, interacting with it and with each other [...]". As shown above, this definition is beneficial and in accordance with the ecosophy of this research and its norms no. 1: Valuing Living and no. 2: Wellbeing, since it implies stability and harmony among the distinct species in the ecological system. Another beneficial definition of "ecosystem" agreeing with the ecosophy and these two norms is the one from Oxford's *A Dictionary of Ecology* (Allaby 2015, 136), which further underlines the idea of interconnectedness between the single living beings in the ecosystem and their physical environment. In the ecology and environment texts referenced here, as well as in Text 3, the noun "ecosystem" is also utilised as a premodifier in noun phrases. "Ecosystem services" (Miller and Spoolman 2019, G4) is destructive and anthropocentric, because it runs counter to norms no. 1: Valuing Living and no. 4: Care; "ecosystem engineers", from Text 3, will be investigated below. The various occurrences and uses of the noun "ecosystem" in the context of Text 3 are scrutinised here by distinguishing the stylistic strategies of foregrounding, point of view and metaphor, outlined in the theoretical Sect. 2.2.

The first example of foregrounding can be found in Sentences 2–4. Sentence 2 is an agentless passive clause ("The wood has been chiselled"), viz. a passive clause without an expressed agent introduced by the preposition "by", which conveys that the agent of the material process of "chiselling" is unknown at the moment. In Sentences 3 and 4, the repetition (Leech and Short 2007, 39–41; Mukařovský 1977; see Sect. 2.2.1) of the lexical verb "fell/ed" is deployed to disclose who that agent is not in the former sentence ("a forester did not fell"), and who that agent is in the latter sentence ("They were felled by [...]"). At the end of Sentence 4, in paragraph end-focus, after four full sentences arousing the addressee's curiosity, the agent is uncovered by means of the apposition "one of nature's greatest engineers, the beaver".

The apposition realises a case of semantic deviation (Short 1996, 43; see Sect. 2.2.1) for two reasons. Firstly, in the Oxford English Dictionary and in all the major semantic divisions and senses of the entry "engineer", this noun denotes a person. The most appropriate sense for Text 3, which discusses felling and shaping trees and, in the subsequent sentences, building dams, is sense 3: "A person whose profession is the designing and constructing of works of public utility, such as bridges, roads, canals, railways, harbours, drainage works, etc.". Accordingly, the fact that the engineer in the text is not a person, but an animal, is semantically deviant. Secondly, underlying all the Oxford English Dictionary senses is the nature-culture opposition explored and contradicted in Text 1 (see Sect. 3.2). As observed in this dictionary, an organism dedicated to "the designing and constructing of works of public utility", like dams, must be a person (not an animal), and their work must be a part of culture (not of nature). Nevertheless, via the semantically deviant collocation "nature's greatest engineers", which binds nature and culture, Text 3 reminds its addressee that a nature-culture nexus must be substituted for the nature-culture opposition and, more importantly, that animals have their own culture and can use equally, if not more, specialised knowledge and skills to build and maintain complex equipment.

The semantic deviation "nature's greatest engineers" is reinforced by the noun phrase "ecosystem engineers" (Sentences 12 and 18). In the discipline of ecology, the term "ecosystem engineers" was coined in 1994 by Clive G. Jones and his co-authors to refer to "organisms that directly or indirectly modulate the availability of resources to other species, by causing physical state changes in biotic or abiotic materials. In so doing they modify, maintain and create habitats" (Jones et al. 1994, 373). These ecologists add that beavers are familiar examples of ecosystem engineers. The term is not recorded in the Oxford English Dictionary, or included in the ecology and environment dictionaries and textbooks I consulted; therefore, it can be safely hypothesised that it is relatively unknown to the general public.

As such, "ecosystem engineers" is, or sounds like, another semantically deviant collocation, since engineers are commonly people, whereas in Text 3 they are beavers. The noun phrase is especially salient, given that it is the only phrase to be present twice in the text, once in a prominent heading position (Sentence 12). According to the Oxford English Dictionary, sense 2.a, the noun "engineer" can be frequently found with a distinguishing word, such as "chemical, electrical, genetic, mechanical, software, systems". Therefore, the distinguishing word "ecosystem" is not among those most often employed to premodify the noun "engineer": this is why the collocation "ecosystem engineers" is an example of semantic deviation. The Oxford English Dictionary evidence is supported by that supplied by the ecology and environment dictionaries and textbooks I consulted, analysed in Sect. 5.1: "ecosystem" as a distinguishing word is only utilised to premodify the noun "services" in the destructive noun phrase "ecosystem services". Most of all, as in "nature's greatest engineers", "ecosystem" evokes the concept of nature, while "engineers" the seemingly opposite notion of culture. However, as "ecosystem engineers" are not people but beavers, the nature-culture opposition dissolves and the existence of the nature-culture nexus is affirmed.

The device of repetition also occurs in Sentences 5–23; what is repeated here is not words, but syntactic structures. More precisely, these sentences include eleven two-part coordinated structures at several syntactic levels, for instance:

1. Noun phrases "a large semi-aquatic mammal and one of the largest rodents in the world" (Sentence 5);
2. Noun phrases "Their thick, waterproof fur and broad, leathery tail" (Sentence 6);
3. Coordinated adjectives "thick, waterproof" and "broad, leathery" (Sentence 6);
4. Noun phrases "tree felling and dam building" (Sentence 7);
5. Noun phrases "their fur and scent glands" (Sentence 9);
6. Lexical verbs "modify and shape" (Sentence 14);
7. Noun phrases "complex pools and riffles" (Sentence 15);
8. Noun phrases "deeper water and shallow, fast-flowing areas" (Sentence 15);

9. Finite clause "are often referred to as 'ecosystem engineers'" and elliptical finite clause "known as a keystone species" (Sentence 18);
10. Finite clauses "increase water storage and slow the flow of water downstream" (Sentence 21);
11. Nouns "branches and vegetation" (Sentence 23).

Through syntactic repetition, these eleven structures depict beavers from a naturalistic standpoint by offering a wide variety of details. This long and thorough list makes the addressee of Text 3 look at beavers with wonder at their bodies, skills and achievements (although this cannot be directly inferred from the text, beavers are also admired for their work ethic: this admiration is lexicalised in the verb "to beaver away", which means to work very hard for a long time [OED Online 2022]).

Another stylistic strategy typifying Text 3 is point of view, to be more precise ideological point of view (Simpson 1993, 11–19, 27–39, 96–101; summarised in Neary 2014, 177–178; see Sect. 2.2.2). This is communicated throughout the text via value-laden words. The negatively value-laden string "heavily hunted" (Sentence 9) and adjective "extinct" (Sentence 10) refer to beavers' history in Britain prior to the species reintroduction projects reported in the second part of the article following Text 3. These words mentioning unpleasant facts make the addressee feel sympathy for beavers and sustain these projects. Text 3 also includes many positively value-laden noun and adjective phrases describing beavers, their features and their functions in the ecosystem. The most pleasant in the context of the text are the following:

1. "one of nature's greatest engineers" (Sentence 4);
2. "one of the largest rodents in the world" (Sentence 5);
3. "excellent swimmers" (Sentence 6);
4. "huge orange teeth hardened with iron" (Sentence 7);
5. "These charismatic creatures" (Sentence 8);
6. "an important role" and "complex wetland ecosystems" (Sentence 13);
7. "complex pools and riffles" and "important for a range of aquatic life" (Sentence 15);
8. "great for breeding birds" (Sentence 16);

9. "an abundance of invertebrates" (Sentence 17);
10. "a keystone species" (Sentence 18);
11. "an entire ecosystem" (Sentence 19);
12. "a real feat of engineering in themselves" (Sentence 22).

These twelve phrases characterise beavers as an impressive, if not unique, species. For instance, the ecosystems "engineered" by beavers are "entire" and "complex", and "complex" also describes the "pools and riffles" created by their damming. Being "huge" or "orange" or "hardened with iron" are not necessarily agreeable characteristics in any context; nevertheless, in Text 3, these adjectives and this adjective phrase become striking, because they premodify and postmodify the noun "teeth", and because of what beavers can accomplish with their teeth. These noun and adjective phrases explicitly recall the Oxford English Dictionary definition of the marker word "ecosystem" examined in Sect. 5.1, which is founded on the idea of interaction between animal and vegetable organisms. In fact, beavers interact with "a range of aquatic life", "birds" and "an abundance of invertebrates", and are a pivotal factor in these organisms' lives and wellbeing. As a result, ideological point of view and positive appraisal are employed in Text 3 to urge the addressee to appreciate beavers and their diverse traits and to value the ecological work they do.

The foregrounding practice of repetition in Sentences 2–4, studied above, combines with the technique of spatial and psychological point of view (Simpson 1993, 11–19, 27–39, 96–101; summarised in Neary 2014, 177–178; see Sect. 2.2.2) in the entire opening paragraph composed of Sentences 1–4. The opening non-finite clause "Walking along the water's edge" (Sentence 1) lacks a grammatical subject to relay who the agent of the material process of "Walking" is: this unexpressed subject is the third-person narrator of Text 3. While walking, the narrator positions the addressee with them by means of the locative expression "along the water's edge" and progressively discovers the items in the landscape they are exploring. Firstly, they visually perceive that "small tree stumps dot the area" (Sentence 1); secondly, that "The wood has been chiselled" (Sentence 2); lastly, that there are "Webbed hind footprints in the mud" (Sentence 3). Furthermore, it is the narrator who assesses that the shape of the wood appears to be "a pencil-like point, as if sharpened with a knife" (Sentence 2), who reads the "sign" of more-than-human activity (Sentence 3), and who finally concludes that the agent of these more-than-human processes is the beaver (Sentence 4). This effect is called "psychological sequencing": "Information which would normally be supplied straight away is withheld for a moment. This prompts us to infer that the narrator must have departed from the norm for some reason, to represent the narrator's perceptions" (Short 1996, 275). In Text 3, this mechanism is also deployed to portray the narrator's line of reasoning, to provide the addressee with more and more clues as to who the "landscape engineer" is, and to get and hold the addressee's attention.

The noun phrase "a pencil-like point" (Sentence 2), the clause "as if sharpened with a knife" (Sentence 2) and the noun phrase "another sign that a forester did not fell these trees" are linguistic indicators of the narrator's psychological point of view. More precisely, they express perception modality, namely the narrator's assurance in the veracity of their propositions as based on their visual and similar powers of perception (for point of view and modality, see Simpson 1993, 43–47; summarised in Neary 2014, 181–182, and in Sect. 2.2.2). Underlying Paul Simpson's theory of modality and psychological point of view is the one elaborated by Roger Fowler. The latter scholar (Fowler 1986, 142) points out that words such as "as if" and "sign" and comparisons such as "pencil-like" signal estrangement and an act of interpretation: the narrator is trying to reconstruct what has happened in the landscape they are going through by relying on external observation. These words of estrangement express a medium value of narrator commitment to the propositions depicting the "beaver-engineered" landscape.

In Text 3, apart from the opening paragraph consisting of Sentences 1–4, psychological point of view is indicated through modality towards the end of the text, in Sentences 18–21. Here, four modal operators—the adverb "often" (Sentence 18), the auxiliary verb "can" (twice, Sentences 19 and 20) and the adverb "potentially" (Sentence 20)—all suggest epistemic modality; that is to say, they evoke the addresser's cognition, beliefs and knowledge, and whether they are convinced of the truthfulness of the modalised propositions. As shown in Sects. 3.2.1 and 4.2.1, in line with the Hallidayan perspective on modality (Halliday 2014, 176–183), "often" is incorporated into the epistemic modal system, given that it conveys a value from the scale of usuality and oftenness "sometimes/usually/always", specifically a medium value. A medium value of addresser commitment is also relayed by the two instances of "can", while "potentially" communicates weak commitment.

In short, Text 3 is framed by perception modality at the beginning, and by epistemic modality towards the end, both with medium values and medium addresser commitment (with the exception of low-value "potentially"). This modality pattern generally signals negative modal shading (for modal shading, see Simpson 1993, 50–69; summarised in Neary 2014, 183–187, and in Sect. 2.2.2), i.e. the addresser's uncertainty about the accuracy of what they are psychologically judging. However, negative modal shading does not seem to typify Text 3. As previously shown, Sentences 1–4, with the three words of estrangement, contain an example of psychological sequencing which, in the unmodalised Sentence 4, culminates in the statement that the "landscape engineer" is the beaver. Moreover, in Sentences 18–21, three of the four epistemic modal operators psychologically evaluate the assertions that beavers "help to restore an entire ecosystem" (Sentence 19), that "Their return […] benefit[s] people too" (Sentence 20), and that beaver dams "reduc[e] the impact of flooding in the surrounding area" (Sentence 21). These assertions are not about present facts which can be measured and quantified accurately; they are about future events very likely, but not certain, to take place, considering the many variables in the ecosystem which can affect them.

What is encompassed by perception modality opening Text 3 and epistemic modality closing it? As shown above, Sentences 4–17, plus the concluding Sentences 22–23, portray beavers and their principal aspects, behaviours and accomplishments. None of these sixteen sentences features modal operators; consequently, they are unmodalised categorical assertions, and the depiction of beavers they furnish is presented as truthful and not psychologically assessed by the addresser. This constitutes neutral modal shading, which is the highest possible level of addresser commitment to what they argue. In Text 3, intermingling neutral modal shading with the lack of negative modal shading investigated above results in a self-confident addresser and a descriptive text abounding in unquestionable factual evidence about beavers, which cannot but convince the addressee that these animals are indeed remarkable.

Text 3 is hence rich in objective facts, indicated by the stylistic devices of foregrounding and point of view scrutinised above; by the same token, it is not rich in metaphors or figurative language. Nevertheless, a conceptual metaphor (Lakoff and Johnson 2003; see Sect. 2.2.3) is triggered by the semantically deviant noun phrase "ecosystem engineers" (Sentences 12 and 18). The conceptual metaphor can be revealed and explored by applying Steen's (2016; see Sect. 2.2.3) three-dimensional model of metaphor, as follows:

1. At the level of expression, or metaphor in language, the conceptual metaphor is prompted by the noun phrase "ecosystem engineers". The Oxford English Dictionary defines the noun "engineer" as "[…] a person who uses specialized knowledge or skills to design, build, and maintain complicated equipment, systems, processes, etc. […]" (sense 2.a), and as "A person whose profession is the designing and constructing of works of public utility […]" (sense 3). Accordingly, taking these two senses into consideration, the noun phrase "ecosystem engineers" can be glossed as "the agents designing, building, and maintaining an ecosystem", and as "the agents designing and constructing an ecosystem".
2. At the level of conceptualisation, or metaphor in thought, the two glosses of the noun phrase "ecosystem engineers" founded on the Oxford English Dictionary senses elicit the target ECOSYSTEM and the source COMPLICATED EQUIPMENT, SYSTEM, PROCESS AND/OR WORK OF PUBLIC UTILITY. This results in the conceptual metaphor ECOSYSTEM IS COMPLICATED EQUIPMENT, SYSTEM, PROCESS AND/OR WORK OF PUBLIC UTILITY.
3. At the level of communication, or metaphor in interaction, ECOSYSTEM IS COMPLICATED EQUIPMENT, SYSTEM, PROCESS AND/OR WORK OF PUBLIC UTILITY is not set down in the three databases Master Metaphor List (Lakoff and Cognitive Linguistics Group 1991), Metalude (Goatly and LLE Project 2002–2005) and

> MetaNet Metaphor Wiki (David et al. 2016–2018). However, with a view to understanding whether this is a creative metaphor with a deliberate figurative function, the communication and interaction between the addresser and the addressee of Text 3 must be taken into account. Text 3 is not a peer-reviewed article written by an ecologist and addressed to their disciplinary and professional community. In that context, both the addresser and the addressee would be acquainted with the term "ecosystem engineers" activating this conceptual metaphor; therefore, it would not be thought of as a creative metaphor. On the contrary, Text 3 is published on Forestry England's website and addressed and easily accessible to people with ecological concerns, if not to the public at large. In this different context, the addresser is very likely to be aware that "ecosystem engineers" is a technical term, but the addressee is not; as a result, the latter may consider it to be a case of creative use of language, and so the conceptual metaphor it triggers.

What is instead reported in Metalude, one of the three metaphor databases, is two root analogies based on the lexical items "genetic engineering" and "social engineering", which have the literal meaning of "design and building of machines[,] engines and electrical equipment or infra". "Genetic engineering" underlies the root analogy HUMAN IS MACHINE/APPLIANCE; it has the metaphorical meaning of "changing the structure of genes to 'improve' or make more healthy", which is exemplified by the sentence "We need more legal safeguards to protect us against genetic engineering". "Social engineering" underpins the root analogy ORGANISATION/SYSTEM IS MACHINE; its metaphorical meaning is "artificial control of or change to the groupings within society", which is illustrated by the example "The campaign to persuade graduates to have more children is a blatant piece of social engineering". Consequently, the lemma "engineer" and its word forms are cognitively associated with machines and appliances; as can be inferred from the example sentences, they acquire the negative connotation of artificial radical change to natural organisms, which may be dangerous and threatening.

Both the conceptual metaphor ECOSYSTEM IS COMPLICATED EQUIPMENT, SYSTEM, PROCESS AND/OR WORK OF PUBLIC UTILITY and the two root analogies analysed above remind the addressee of the conceptual metaphor NATURE IS A MACHINE, which is destructive and diverges from the ecosophy of this research. Firstly, this conceptual metaphor depicts nature and the ecosystem as a machine composed of a collection of pieces which may be repaired or replaced individually, without having to consider the entire system and its interconnectedness. Secondly, the metaphor does not acknowledge or praise the animal and vegetable species living in and contributing to nature and the ecosystem: they just become mere components (Stibbe 2021, 64–65). Hence, Text 3 must be rid of all these negative connotations and links to destructive metaphors and root analogies, and made more beneficial. In order to do so, although the noun phrase "ecosystem engineers" is semantically deviant and, as such, foregrounded, it must be rephrased following the ecosophy. This will be done in the evaluative Sect. 5.2.2.

5.2.2 Summary and Evaluation

After the ecostylistic study of Text 3, drawn from Forestry England's website, including the marker word "ecosystem", and the investigation of the stylistic practices of foregrounding, point of view and metaphor figuring in the text, Text 3 and these practices will be summarised and judged. More precisely, this section evaluates the value-system of Text 3, compares it with the ecosophy of this research and its norms, and shows whether the text and its practices are actually beneficial.

As mentioned in Sect. 5.2.1, the concept of ecosystem in Text 3 is founded on the beneficial definitions of this marker word provided by the Oxford English Dictionary and by Oxford's *A Dictionary of Ecology* (Allaby 2015, 136): an ecosystem consists of all the animals and plants living in an environment and is characterised by their mutual exchange and connection. In Text 3, prominence is given to beavers and to the part they play in fashioning and replenishing ecosystems, in keeping them healthy and in helping the more-than-human world living there to thrive. These definitions, together with Text 3 they underlie, entail balance and cooperation among the various animal and vegetable organisms in the ecosystem; as a result, they align with norms no. 1: Valuing

Living and no. 2: Wellbeing, which concern these ecosystemic principles. Consequently, equilibrium and collaboration, as exemplified by beavers and their ecological work, are beneficial notions to be disseminated among the general public. Beavers, their qualities and their tasks are portrayed and underscored in Text 3 by means of the stylistic techniques of foregrounding, point of view and metaphor; in several sentences, two techniques co-occur to highlight central ideas in the text and to make them more conspicuous to the addressee.

In Sentences 2–4, the foregrounding mechanisms of the agentless passive clause "The wood has been chiselled", the repetition of the lexical verb "fell/ed" and the semantically deviant apposition "one of nature's greatest engineers, the beaver" all have the function of stoking the addressee's interest in the more-than-human world in general and in these animals in particular. This interest is also roused by the semantic deviation "ecosystem engineers" in Sentences 12 and 18. If the addressee of Text 3 is by any chance persuaded that only we humans can design and produce objects by physical and mental work, the text supersedes that belief by firmly contending that nonhuman species can develop and maintain complex systems with their knowledge and abilities. These stylistic strategies also have the role of reminding the addressee that nature ("nature", "ecosystem") is not the opposite of culture ("engineers"), but that these two concepts are intimately related in a nexus, as posited by Text 1. These ideas are beneficial and in accordance with the ecosophy of this research and its norm no. 1: Valuing Living: it maintains that all animal and vegetable organisms must be extolled and are deserving of high consideration, and that the ecosphere must be conceived of as a single unit.

Many instances of the foregrounding practice of repetition are present in most of Text 3, especially in Sentences 5–23; to be more precise, the text shows the repetition of syntactic structures at distinct levels, in eleven two-part coordinated structures describing beavers, such as the noun phrases "a large semi-aquatic mammal and one of the largest rodents in the world" (Sentence 5). Repetition frequently co-occurs with the stylistic device of ideological point of view, particularly in Sentences 4–22. The description of beavers is made via twelve positively value-laden noun and adjective phrases depicting them as extremely notable animals, like "huge orange teeth hardened with iron" (Sentence 7). Moreover, in Sentences 4–17 and 22–23, repetition and ideological point of view also co-occur with psychological point of view, more precisely with the lack of it. These sixteen sentences are unmodalised categorical assertions with neutral modal shading, which convey an accurate picture of beavers not psychologically evaluated by the addresser. Another picture is also conveyed by unmodalised categorical assertions: that of a self-assured addresser of a detailed text replete with undeniable facts concerning beavers.

Given the effect of repetition and of ideological and psychological point of view on the addressee of Text 3, after reading this extensive and detailed description, the addressee cannot but look at beavers with admiration for their bodies, talents and accomplishments in the ecosystem. For the feeling of wonder they generate, these stylistic techniques are beneficial and agree with the ecosophy and its norm no. 1: Valuing Living. In addition, Text 3 portrays the interplay between beavers and the more-than-human life in the ecosystem they contribute to building; what they also contribute to is the wellbeing of all the animal and vegetable species living there. This is in line with norm no. 2: Wellbeing: thanks to their interplay, all the species in the ecosystem thrive and enjoy good health, as well as physical and psychological wellbeing.

In Sentences 8–11, the description of beavers offers information about their history in Britain and comprises two negatively value-laden items: the string "heavily hunted" (Sentence 9) and the adjective "extinct" (Sentence 10). The former is a constituent in the agentless passive clause "beavers were heavily hunted […]", whose omitted agent is us humans. As asserted by Quirk et al. (1985, 165), "This omission [of the agent] occurs especially when the agent is irrelevant or unknown […] or when the agent is left out as redundant". In the ecological discourse of Text 3, the agent of the material process of heavily hunting a species, thus causing its extinction, cannot be held to be irrelevant; the agent cannot be unknown either, since the hunters after "their [beavers'] fur and scent glands" (Sentence 9) can only be us humans, not other organisms. Hence, the addresser of the text is likely not to have openly hinted at us humans as the agent hunting a species to extinction, given that they considered the agent to be redundant and predictable from knowledge of the context.

Accordingly, Text 3 does not fully follow the ecosophy and its norms no. 2: Wellbeing and no. 4: Care. The former claims that we humans must fight extinction by protecting all animals and plants and by defending their abundance and variety; the latter norm contains the notions of empathy and regret and notes that we must be aware of our negative impact on and damage to the more-than-human world, and work to limit that impact and that damage. Therefore, if Text 3 had referred straightforwardly to us humans as the agent in Sentences 9 and 10, it would have been more beneficial and would have better aligned with the ecosophy. A possible rewriting of these sentences can be: "we humans heavily hunted beavers for their fur and scent glands that produce a substance called castoreum. Due to our relentless hunting activities, by the beginning of the 16th century the beaver became extinct in the country". In this rewriting, not only is the human agent, along with our human blame, introduced into this sequence of Text 3; in Sentence 9, now in the active voice, the agent realises the salient function of subject, while in Sentence 10 "our relentless hunting activities" are in marked thematic position, and their association with beavers' extinction is made explicit.

With regard to the stylistic mechanism of spatial and psychological point of view, Text 3 utilises the effect of psychological sequencing in the entire opening paragraph (Sentences 1–4) and the words of estrangement "(pencil-)like", "as if" (Sentence 2) and "sign" (Sentence 3), which are operators of perception modality. These strategies arouse the addressee's curiosity about beavers and encourage them to read on so as to find out more about this species. As a result, these are beneficial devices in accordance with the ecosophy and its norm no. 1: Valuing Living: any animal or vegetable organism has inherent value, is an end in itself and, as such, deserves our interest and to be known and regarded very highly.

In addition to perception modality, psychological point of view in Text 3 is also conveyed through four linguistic indicators of epistemic modality, which is deployed in Sentences 18–21 to assess events to come in the future. Lower degrees of perception modality and of epistemic modality normally bring about negative modal shading, with an addresser doubtful about the factuality of the propositions they are psychologically appraising; nevertheless, as shown in Sect. 5.2.1, this is not the case with Text 3. One of these propositions is "Their [beavers'] return can benefit people too" (Sentence 20). On the one hand, the ecosophy, in its norm no. 2: Wellbeing, observes that fulfilling our human vital and spiritual needs and prioritising our own wellbeing are fundamental conditions for living well. On the other hand, the same norm points out that the wellbeing of all animals and plants must be given special consideration. Consequently, the idea of beavers' return also benefitting us humans is ambivalent (see Sect. 2.1.3) and does not go strongly against the ecosophy.

However, a text considering the concept of ecosystem must go beyond our human needs and benefits, even the vital ones, and must mention both those regarding the human world and those regarding the more-than-human world. Hence, Sentence 20 can be rephrased in a more beneficial and less ambivalent way as: "Their return can benefit all the animal and vegetable species living in their ecosystem, and even people not living there". This rephrasing brings all organisms, human and nonhuman, to the same level; furthermore, the conventional implicature of unexpectedness prompted by the adverb "even", which replaces the original adverb "too", concurs in describing beavers as important animals able to influence and improve ecosystems they do not live in.

The notion of ecosystem, in Text 3, is also represented by means of the conceptual metaphor ECOSYSTEM IS COMPLICATED EQUIPMENT, SYSTEM, PROCESS AND/OR WORK OF PUBLIC UTILITY. Activated by the noun phrase "ecosystem engineers" (Sentences 12 and 18), this metaphor is destructive when compared with the ecosophy: it is semantically and conceptually connected with the destructive root analogies HUMAN IS MACHINE/APPLIANCE and ORGANISATION/SYSTEM IS MACHINE, and with the destructive metaphor NATURE IS A MACHINE. More precisely, the conceptual metaphor runs counter to norms no. 1: Valuing Living and no. 7: Adaptation. Firstly, all lives in the ecosystem must be valued because they have worth in themselves, not because they are made equal to a manufactured machine or appliance. Secondly, the metaphor may erroneously lead the addressee to believe that the ecosystem is as resilient as an insentient and inanimate machine, and that we humans do not need to develop new economic and social systems with a view to safeguarding it.

Accordingly, the conceptual metaphor ECOSYSTEM IS COMPLICATED EQUIPMENT, SYSTEM, PROCESS AND/OR WORK OF PUBLIC UTILITY must be reworded in a beneficial way. Its source COMPLICATED EQUIPMENT, SYSTEM, PROCESS AND/OR WORK OF PUBLIC UTILITY, as it stands, is not suitable to portray the target ECOSYSTEM in a beneficial metaphor; therefore, the noun phrase "ecosystem engineers" is not suitable to portray beavers in a beneficial text. When the linguistic and naturalistic data from Text 3 are borne in mind, the association between the target ECOSYSTEM and its source is more accurately represented not with a metaphor, but with the frame A Beaver Ecosystem is A Complicated System with Complicated Processes Including Works of Ecosystemic Utility (for frames, see Tannen 1993. In accordance with common typographical conventions, the initials of frames are Capitalised, and metaphors, targets and sources are in SMALL CAPITALS). Contrary to the conceptual metaphor, this frame does not diverge from the ecosophy and its norms no. 1: Valuing Living and no. 7: Adaptation.

Nevertheless, as shown in Sect. 5.2.1, the noun phrase "ecosystem engineers" is prominent and holds the addressee's attention, as it is semantically deviant and metaphorical; as a result, an equally deviant and metaphorical beneficial alternative must be proposed. A possible counterpart can be the noun phrase "ecosystem carpenters": it openly hints at beavers' most important deed, viz. building dams, and elicits the conceptual metaphor ECOSYSTEM IS WOODWORK. This is a creative metaphor, with a deliberate figurative function, not listed in the three metaphor databases referenced in this book. This creative metaphor has at least three advantages. The first is that it fully agrees with the ecosophy. The second advantage of this conceptual metaphor is that it does not cognitively link the ecosystem to machines or appliances, nor to their negative connotation of artificiality and potential danger scrutinised in Sect. 5.2.1. The third advantage of the metaphor is that it substitutes the natural material of wood for manufactured machines and appliances: this cancels the idea of threat and turns the negative connotation of artificiality into the positive connotation of being in accordance with nature. In addition, both the noun phrase "ecosystem carpenters" and the conceptual metaphor ECOSYSTEM IS WOODWORK emphasise beavers' craftsmanship and astonishing skills. As demonstrated by the ecostylistic analysis of Text 3 and by its comparison with the ecosophy of this research, Forestry England's environmental website encourages us humans to change our mainstream mindset and to remember that each animal and vegetable species in the ecosystem is as unique as we think we are and has the same intrinsic value as we do.

References

Allaby, M. 2015 [1994]. *A Dictionary of Ecology*, 5th ed. Oxford: Oxford University Press.

Collin, P.H. 2011 [1985]. *Dictionary of Environment & Ecology*, 5th ed. London: Bloomsbury [electronic edition].

David, O., G. Lakoff et al. 2016–2018. *MetaNet*. Berkeley: University of California. https://metaphor.icsi.berkeley.edu/pub/en. Last accessed April 2022.

Fowler, R. 1986. *Linguistic Criticism*. Oxford: Oxford University Press.

Goatly, A., and LLE Project. 2002–2005. *Metalude*. Hong Kong: Lingnan University. http://www.ln.edu.hk/lle/cwd/project01/web/home.html. Last accessed April 2022.

Halliday, M.A.K. 2014 [1985]. *Halliday's Introduction to Functional Grammar*, revised by C.M.I.M. Matthiessen, 4th ed. London and New York: Routledge.

Jones, C.G., J.H. Lawton, and M. Shachak. 1994. Organisms as ecosystem engineers. *Oikos* 69 (3): 373–386.

Lakoff, G., and Cognitive Linguistics Group. 1991. *Master Metaphor List*. Berkeley: University of California. http://araw.mede.uic.edu/~alansz/metaphor/METAPHORLIST.pdf. Last accessed April 2022.

Lakoff, G., and M. Johnson. 2003 [1980]. *Metaphors We Live By: With a New Afterword*, 2nd edition. Chicago: University of Chicago Press.

Larson, B. 2011. *Metaphors for Environmental Sustainability: Redefining Our Relationship with Nature*. New Haven, CT: Yale University Press.

Leech, G., and M. Short. 2007 [1981]. *Style in Fiction: A Linguistic Introduction to English Fictional Prose*, 2nd ed. Harlow: Longman.

Miller, G.T., and S.E. Spoolman. 2019 [2007]. *Essentials of Ecology*, 8th ed. Belmont, CA: Cengage.

Mukařovský, J. 1977. On poetic language. In *The Word and Verbal Art: Selected Essays*, ed. J. Burbank and P. Steiner, 1–64. New Haven, CT and London: Yale University Press.

Myerson, G., and Y. Rydin. 1996. *The Language of Environment: A New Rhetoric*. London and New York: Routledge.

Neary, C. 2014. Stylistics, point of view and modality. In *The Routledge Handbook of Stylistics*, ed. M. Burke, 175–190. London and New York: Routledge.

Oxford English Dictionary (OED) Online. 2022. Oxford: Oxford University Press. https://www.oed.com/. Last accessed April 2022.

Quirk, R., S. Greenbaum, G. Leech, and J. Svartvik. 1985. *A Comprehensive Grammar of the English Language*. Harlow: Longman.

Short, M. 1996. *Exploring the Language of Poems, Plays and Prose*. Harlow: Longman.

Simpson, P. 1993. *Language, Ideology and Point of View*. London and New York: Routledge.

Steen, G. 2016. Metaphor: Metaphor and style through genre, with illustrations from Carol Ann Duffy's *Rapture*. In *The Bloomsbury Companion to Stylistics*, ed. V. Sotirova, 308–324. London: Bloomsbury.

Stibbe, A. 2021 [2015]. *Ecolinguistics: Language, Ecology and the Stories We Live By*, 2nd ed. London and New York: Routledge.

Tannen, D., ed. 1993. *Framing in Discourse*. Oxford: Oxford University Press.

6

Ecology

6.1 Definition and Evaluation

The fourth of the five marker words (Myerson and Rydin 1996, 6; see Sect. 1.1) analysed in this book is "ecology". The senses of this noun provided by the Oxford English Dictionary (OED Online 2022) are the following:

> **1.a.** The branch of biology that deals with the relationships between living organisms and their environment. Also: the relationships themselves, esp. those *of* a specified organism. See also BIOECOLOGY *n.*
> **1.b.** Chiefly *Sociology*. The study of the relationships between people, social groups, and their environment; (also) the system of such relationships in an area of human settlement. Frequently with modifying word, as *cultural ecology*, *social ecology*, *urban ecology*.
> **1.c.** In extended use: the interrelationship between any system and its environment; the product of this.
> **2.** The study of or concern for the effect of human activity on the environment; advocacy of restrictions on industrial and agricultural

D. F. Virdis, *Ecological Stylistics*,
https://doi.org/10.1007/978-3-031-10658-3_6

> development as a political movement; (also) a political movement dedicated to this. Cf. *deep ecology n.* at DEEP *adj.* Compounds 2. (OED Online 2022)

In sense 1.a, the noun "ecology" is defined as the discipline exploring the associations between living organisms and their environment; consequently, the object of study of this field is both human life and more-than-human life, with no distinction. Sense 1.b, concentrating on our human world, describes the three noun phrases "cultural ecology", "social ecology" and "urban ecology" as being principally utilised in the area of sociology; "urban ecology" is also well attested in the discipline of ecology proper with the meaning of "The study of the ecology of organisms in the context of the urban environment" (Smith and Smith 2015, 671; see also Editors of Salem Press 2004, 726). In sense 2, the noun phrase "the effect of human activity on the environment" triggers the existential presupposition that such an effect exists, but does not state whether it is beneficial or adverse. The noun "study", being value-neutral, does not contribute to explaining this; however, the noun "concern", with its value-laden senses of "a subject that excites one's interest, attention, or care" and even "anxiety", appears to suggest that our human effect on the environment is undesirable, to say the least. This reading is reinforced by the noun phrase "advocacy of restrictions" on our human activity, so that nature can be safeguarded: if that activity were not malign, those restrictions would not be advocated. Hence, as the definition reminds us, "a political movement" or, rather, several political movements and projects worldwide were created and developed in order to conserve the animal and vegetable species on the planet (for an outline of "deep ecology", referred to in the sense 2, see Sect. 2.1.2).

The ecosophy of this research (see Sect. 1.3), in its norm no. 1: Valuing Living, respects all human, animal and vegetable organisms alike. Furthermore, norm no. 4: Care champions solicitude for and serious attention to animals and plants, in short the "concern" for more-than-human life and the "advocacy of restrictions" to preserve it mentioned in the Oxford English Dictionary and in its sense 2. It follows from such "concern" and "advocacy of restrictions" that the study and movement in the field of ecology are aware of norms no. 3: Now and the

Future and no. 5: Environmental Limits. In other words, ecology must be concerned with protecting the natural world not only at present, but also in the decades to come, and must suggest environmental restrictions with a view to the future and to taking environmental limits into consideration. Accordingly, the aims and scope of ecology as depicted in the definition are beneficial (see Sect. 2.1.2) when appraised against the ecosophy. The text would have been more beneficial if it had made explicit what it evokes implicitly, viz. that our human effect on the environment is drastic, as we are using its natural materials to excess (norm no. 5: Environmental Limits).

In a future update of the Oxford English Dictionary, a possible beneficial version of the entry "ecology" could clarify that the ecological movement is not a single philosophical group or an undivided political organisation, but that this is an umbrella term deployed to indicate an extensive variety of movements, from those preserving the status quo to those promoting sweeping reforms (see Sect. 1.3 and, among others, Stibbe 2021, 13–14). The same possible beneficial version could also clarify that the thrust of the most radical of these political movements is not merely the "advocacy of restrictions on industrial and agricultural development", but more substantial and far-reaching changes in our lifestyles, in accordance with the ecosophy and its norms no. 6: Social Justice and no. 7: Adaptation.

All the recent ecology and environment dictionaries and textbooks in their latest editions or reprints I consulted include definitions of the noun "ecology". Most of them are analogous to sense 1.a in the Oxford English Dictionary in length and content, but two offer additional details worth analysing in the context of this book. The definition in Routledge's *The Environment Dictionary* (Kemp 2002, 116) states that the German biologist Ernst Haeckel is usually credited with having coined the term "ecology" in 1866, and that ecological examination may be undertaken at various biological levels, listing a number of subdivisions and classifications of the area. The closing sentence reads:

> Popular usage of the term tends to be loose, and is commonly applied to environmental problems that have a human origin. (Kemp 2002, 116)

This roughly corresponds to sense 2 in the Oxford English Dictionary, with a notable difference. As previously shown, this sense only indirectly conveys how dramatic our human influence on the earth is; by contrast, the closing sentence from Routledge's *The Environment Dictionary* employs the noun phrase "environmental problems that have a human origin". By means of the noun "problems", the sentence relays the existence of difficult matters which exercise the mind, are seemingly insoluble and often long-standing, and affect a specified group or nation; moreover, and more importantly, these difficult matters are described as being initiated by us humans, which makes the cause of our ecological problems overt. The scientific facts and figures underpinning the ecosophy in general and its norm no. 5: Environmental Limits in particular have shown that we humans are overusing nature and natural materials; as a result, this definition, which points this out, is beneficial and in line with the ecosophy.

The second definition of the noun "ecology" supplying more information than the ones in the other ecology and environment texts figures in the textbook *Ecology Basics* (Editors of Salem Press 2004, 171), and introduces the chapter devoted to this subject:

> [1] Ecology is the study of the relationships of organisms to their environments. [2] By examining those relationships in natural ecosystems, ecologists can discover principles that help humankind understand its own role on the planet. (Editors of Salem Press 2004, 171)

This definition is divided into two sentences. While Sentence 1 communicates the same details as those expressed by the Oxford English Dictionary in sense 1.a, Sentence 2 refers to the concept of "natural ecosystems". Although the meaning of this term is not stated in the textbook or in its glossary, it can be inferred from its occurrences in the text: a "natural ecosystem" is an ecological system that has not been artificially made or constructed or affected by our human intervention or pressures, and is the opposite of, for instance, an agricultural ecosystem (Editors of Salem Press 2004, ix, 687). Sentence 2 also mentions "principles that help humankind understand its own role on the planet", a noun phrase triggering the existential presupposition that such principles exist and can be revealed, and presuming that it is our duty as humans to make sense of the part we play in nature. The definition closes without clarifying what our part must be in the opinion of the textbook authors, namely whether they reckon that our role should be domineering or more balanced, therefore whether the principles to be discovered are beneficial or destructive (see Sect. 2.1.4).

This clarifying information is provided later on in the chapter about ecology introduced by the definition of this term, in the paragraph headed "Human impacts and applications" in general and in the following sequence in particular:

> [1] People historically have viewed nature as an adversary. [2] The "conquest of nature" has traditionally meant human encroachment on natural ecosystems [...]. [3] [3.1] A major lesson of ecology is that humans are not separate from nature; [3.2] we are constrained by the same principles as are other organisms on the earth. [4] One object of ecology, then, is to learn these principles so that they can be applied to our portion of the earth's ecosystem. (Editors of Salem Press 2004, 176)

Sentences 1–2 concisely portray what is called our "historical" or "traditional" human belief-system about the natural world: it is represented as "an adversary" to be "conquered", whose territories are to be "encroached on", i.e. it is metaphorically conceptualised in aggressive terms through the semantic field of war. Nevertheless, the two sentences also comprise the two lexical verbs "viewed" and "meant", which are two linguistic indicators of the point of view of us humans; as such, they signal that this representation of the environment is not a fact, but is partial and subjective. This is confirmed by the noun "encroachment", which has the negative denotation of intruding usurpingly on another's territory and rights and is utilised to unfavourably describe our human activities pursued at the expense of natural ecosystems (for a definition of "natural ecosystem" in this textbook, see above).

After Sentences 1–2 have suggested and criticised our human perspective on the earth, Sentences 3–4 sum up the scientific, consequently less partial and subjective, standpoint of the discipline of ecology, with its "major lesson" and "object". Sentence 3.1 contains the finite clause "humans are not separate from nature", with the adverb "not" giving the clause negative polarity. In Hallidayan functional grammar, while a positive clause is typically unmarked, a negative clause is typically marked, since it conveys an expectation which is concurrently defied (Halliday 2014, 173–176). In this case, the expectation to be defied is "humans *are* separate from nature", which is an outline of Sentences 1–2.

Sentence 3.2 ("we are [...] the earth") rephrases and gives reasons for the defied expectation by means of several stylistic mechanisms. The noun "humans" in Sentence 3.1 is anaphorically referred to by the personal pronoun "we" in Sentence 3.2: in terms of deixis, this is an example of inclusive we embracing the addresser (here the textbook authors) and the addressee (the textbook readers), thereby involving the latter directly. In addition, the noun phrase "the same principles as are other organisms on the earth" refers back to the *Ecology Basics* definition of "ecology", studied above, and to the "principles that help humankind understand its own role on the planet", indicated there. Sentence 3.2 makes it clear that, according to the textbook authors, the part we play in the natural world must not be overbearing or predominating, but fair and founded on the same physical and biological laws and values that apply to all animal and vegetable species. This is reinforced by the adjective "other", which triggers the lexical presupposition that we humans are living organisms equal to those belonging to the more-than-human world, subject to the same laws and values and with the same rights and responsibilities.

Sentence 4 also defies the expectation that "humans are separate from nature" and puts us humans back in our right perspective. The lexical verb "to learn" links back to the noun "lesson" in Sentence 3.1; these two words from the semantic field of education imply that we humans must be taught and study the findings of the field of ecology. We are hence depicted as pupils in need of learning, and the field as a teacher satisfying this need; that is to say, we are attributed a humble and deferential position. This modest position entails modest pretensions: because our role in the environment must not be domineering but balanced, we are only entitled to "our portion of the earth's ecosystem", not to take advantage of it all. Given the notions and tenets it relays, the sequence under investigation is beneficial when judged against the ecosophy of this research. In particular, underlying the sentences in the sequence are norms no. 1: Valuing Living and no. 5: Environmental Limits: norm no. 1 as the sequence takes the entire ecosphere into consideration, viz. all human, animal and vegetable species; norm no. 5 as the ideas in the sequence involve us humans not using the natural materials in the earth's ecosystem to excess, but deploying those in our portion of it in accordance with the scientific principles of ecology.

6.2 "Ecology" in National Park Service's Words

6.2.1 Ecostylistic Analysis

In the previous section, the Oxford English Dictionary (OED Online 2022) and recent ecology and environment dictionaries and textbooks were quoted to supply a lexicological and naturalistic definition of the marker word "ecology". This section will now examine the use of this marker word in the environet (Myerson and Rydin 1996, 7; see Sect. 1.1), to be more precise in the ecological discourse of National Park Service as relayed by its website (https://www.nps.gov/index.htm) (see Sect. 1.2); in practical terms, the section will study the beneficial ecostylistic techniques linked to the use of this word. Text 4 is the opening paragraph of the article "How Traditional Ecological Knowledge can inform the field of conservation biology". The article, posted on 23 September 2021 in the section "Guest author column" of National Park Service's website, was written by Sarah E. Rinkevich, Assistant Research Scientist at the School of Natural Resources and the Environment, University of Arizona.

Text 4

[1] Indigenous knowledge, also known as Traditional Ecological Knowledge (TEK), is increasingly being applied in the field of wildlife conservation. [2] According to Huntington (1998), TEK offers ecological information and insights relevant to ecological management and research

that cannot be obtained from other sources. [3] For thousands of years, Indigenous peoples have used biological knowledge of their local environment to sustain themselves and to maintain their cultural identity. [4] Indigenous peoples from around the world possess a broad knowledge base of complex ecological systems in their own localities (Gadgil et al. 1993). [5] This information functions within time-tested resource management systems of long-resident peoples. [6] Yet, the involvement of Indigenous people remains rare and Western science often overlooks and disparages these Indigenous systems and associated TEK (Westley and Miller 2003). [7] According to many scholars, TEK parallels the scientific discipline of Ecology because both TEK and Western science share observation and description of the empirical world. [8] Here, I present examples of how TEK has been used to inform the field of Conservation Biology, and use the terms TEK and Indigenous interchangeably.

(Copyright © 2021 National Park Service. "How Traditional Ecological Knowledge can inform the field of conservation biology". By Sarah E. Rinkevich, Ph.D. Last updated September 23, 2021. Available at https://www.nps.gov/subjects/tek/sarah-rinkevich-ph-d-university-of-arizona.htm. No protection is claimed in original U.S. Government works. / No claim to original U.S. Government works. Last accessed April 2022)

[174 words]

In Text 4, the marker word "ecology" occurs once, in Sentence 7: "According to many scholars, TEK parallels the scientific discipline of Ecology because both TEK and Western science share observation and description of the empirical world". The adjective "ecological", which derives from this marker word, can be found four times in three distinct sentences, in the three noun phrases "Traditional Ecological Knowledge (TEK)" (Sentence 1), "ecological information and insights relevant to ecological management and research" (Sentence 2), and "complex ecological systems" (Sentence 4). More importantly, the adjective "ecological" is present seven more times in five sentences (Sentences 1, 2, 6, 7 [twice] and 8 [twice]), as a part of the acronym "TEK". TEK is a main word in Text 4, so much so that the text defines the marker word "ecology" via a comparison with TEK; in other words, a mainstream academic discipline is delineated by drawing a comparison with the empirical experience and wisdom of Indigenous nations and peoples (for the issue of Indigenous nations and peoples in beneficial ecological texts, see also Text 1 and Text 2 and Sects. 3.2 and 4.2). Furthermore, TEK is also a main concept in the ecosophy of this research and in its norms (see Sect. 1.3), especially no. 6: Social Justice and no. 7: Adaptation. For these reasons, Text 4 was chosen for investigation in this book.

Underlying Text 4 and its instances of the marker word "ecology" and of the adjective "ecological", consequently the notion of TEK, are two of the definitions of this marker word provided by the Oxford English Dictionary and by the ecology and environment texts scrutinised in Sect. 6.1: one is offered by the Oxford English Dictionary, the other by *Ecology Basics* (Editors of Salem Press 2004, 171, 176). Text 4 mentions "the scientific discipline of Ecology" (Sentence 7); hence, when the Oxford English Dictionary senses are taken into consideration, the sociological sense 1.b, the extended sense 1.c or the political sense 2 are irrelevant to the text. Also irrelevant is Routledge's *The Environment Dictionary* definition (Kemp 2002, 116), which is similar to the Oxford English Dictionary sense 2. Conversely, sense 1.a delineates the noun "ecology" as "The branch of biology that deals with the relationships between living organisms and their environment. Also: the relationships themselves, esp. those of a specified organism". Accordingly, the aims and scope of the field of ecology as outlined in sense 1.a embrace the scientific analysis of all humans, animals and plants in their environments. This is a beneficial sense in line with the ecosophy and its norm no. 1: Valuing Living, since the entire ecosphere, namely the human world and the more-than-human world alike, is encompassed by the subject and research area of ecology.

An even more beneficial definition is supplied by *Ecology Basics* (Editors of Salem Press 2004, 171, 176), which states that "A major lesson of ecology is that humans are not separate from nature; we are constrained by the same principles as are other organisms on the earth. One object of ecology, then, is to learn these principles so that they can be applied to our portion of the earth's ecosystem". As noted in Sect. 6.1, the "lesson" that we humans are not estranged from the natural world, but are an animal species constituting a part of it, recalls norm no. 1: Valuing Living. It follows that we only have a right to "our portion of the earth's ecosystem", and to no additional portions other organisms are entitled to: this aligns with norm no. 5: Environmental Limits, because we humans must not overuse natural materials, but must only employ those in our portion in accordance with the scientific tenets of ecology (for the idea of "our portion of the earth's ecosystem" in a beneficial ecological text, see also Text 5 and Sect. 7.2). Text 4 shows the connections between the discipline of ecology and TEK and is examined here by pinpointing the stylistic mechanisms introduced in the theoretical Sect. 2.2: foregrounding, point of view and metaphor.

At a syntactic level, Text 4 is typified by long and complex noun phrases, the longest and most complex of which are the following:

1. "Indigenous knowledge, also known as Traditional Ecological Knowledge (TEK)" (Sentence 1);
2. "ecological information and insights relevant to ecological management and research that cannot be obtained from other sources" (Sentence 2);
3. "a broad knowledge base of complex ecological systems in their own localities" (Sentence 4);
4. "time-tested resource management systems of long-resident peoples" (Sentence 5).

These four noun phrases feature articulate premodification (noun phrase "time-tested resource management") and, above all, postmodification (elliptical clause "also known as Traditional Ecological Knowledge (TEK)"; adjective phrase "relevant to ecological management and research"; relative clause "that cannot be obtained from other sources"; prepositional phrase "of complex ecological systems in their own localities"). Articulate premodification and postmodification are characteristic features of academic and scientific writing; in fact, this genre was described as "compressed" and abounding in "compressed noun phrases" by Biber et al. (2021: 324) and by Biber and Gray (2010). In Text 4, the function of compressed noun phrases is to provide the addressee with as many details about TEK as possible; in Sentence 7 ("TEK parallels the scientific discipline of Ecology because [...]"), these details will prove to be the grounds for associating TEK with the academic field of ecology.

The four compressed noun phrases realise examples of foregrounding (Mukařovský 1964; see Sect. 2.2.1), especially of what Emmott and Alexander (2016, 290) called over-specification. They are not syntactically deviant per se (Leech and Short 2007, 39–41; Short 1996, 36–59; see Sect. 2.2.1); however, they are deviant as compared with the noun phrases in the other beneficial texts addressed to the public at large studied in this book, and with the noun phrases in other texts addressed to the same audience on National Park Service's environmental website (see, among others, the ones at https://www.nps.gov/pefo/learn/ecology.htm and https://www.nps.gov/articles/bird-ecology-acad.htm, both including the marker word "ecology", and the one at https://www.nps.gov/subjects/tek/stefani-a-crabtree-phd-with-evan-a-holt-ph-d-student-both-of-utah-state-university.htm, covering TEK and drawn from the same website section as Text 4). As is usually the case with foregrounding, the length and complexity of these noun phrases are likely to increase the amount of time the addressee spends processing the information communicated by the phrases, and the amount of effort the addressee puts into perceiving and experiencing the entire text.

From a syntactic standpoint, a further notable pattern is evident in Text 4: the repetition (Leech and Short 2007, 39–41; Mukařovský 1977; see Sect. 2.2.1) of syntactic structures at various levels, in four two-part coordinated structures with the same or very similar meanings, which therefore strengthen each other's message, like the following:

1. Non-finite clauses "to sustain themselves and to maintain their cultural identity" (Sentence 3);
2. Lexical verbs "overlooks and disparages" (Sentence 6);
3. Noun phrases "these Indigenous systems and associated TEK" (Sentence 6);
4. Noun phrases "observation and description of the empirical world" (Sentence 7).

These repeated syntactic structures clarify and underline primary concepts in Text 4 for the benefit of the addressee, who may be unfamiliar with ecology or with TEK. The two non-finite clauses in Sentence 3 signal that Indigenous peoples utilise TEK to survive and to make their cultural identity survive with them; the fact that the two clauses are coordinated indicates that "sustain[ing] themselves" is analogous to "maintain[ing] their cultural identity", and underscores the importance of Indigenous culture. As expressed by the two lexical verbs "overlooks and disparages" in Sentence 6, TEK is frequently ignored and undervalued by mainstream science. The two noun phrases "these Indigenous systems and associated TEK" in Sentence 6 anaphorically refer to the two compressed noun phrases "a broad knowledge base of complex ecological systems in their own localities" (Sentence 4) and "time-tested resource management systems of long-resident peoples" (Sentence 5); they indicate that these Indigenous systems are so sophisticated and multi-layered that they have evolved their own TEK. As suggested by the two noun phrases "observation and description of the empirical world" in Sentence 7, what ecology and TEK have in common is watching and examining methodically the natural world.

Primary concepts in Text 4 are also conveyed through the stylistic strategy of point of view. As previously mentioned, the text is the opening paragraph of a longer article; Sentence 8 concludes the opening paragraph and introduces the body of the article ("Here, I present examples of how TEK has been used to inform the field of Conservation Biology, and use the terms TEK and Indigenous interchangeably"). By means of the system of deixis and four deictic terms, this sentence relays the spatial and temporal point of view (Simpson 2010, 295; see Sect. 2.2.2) of the addresser of Text 4: she and her addressee share the virtual space of the text (place deictic adverb "here"); the message she is communicating is factual at the time of writing (time deictic lexical verbs "present" and "use"); moreover, she is explicitly engaged in the text and in the message she is communicating (person deictic pronoun "I"). Point of view and deixis are hence used to provide the addresser and the addressee with spatial and temporal common ground, and to directly involve the addressee in the text and the message.

Text 4 also expresses its addresser's ideological point of view (Simpson 1993, 11–19, 27–39, 96–101; summarised in Neary 2014, 177–178; see Sect. 2.2.2) via language choices and lexical selections which are positively value-laden in the context of the text. TEK's academic and scientific validity is given prominence by its having been deployed "around the world" (Sentence 4) and "For thousands of years" (Sentence 3), i.e. by its being "time-tested" (Sentence 5). The relative clause "that cannot be obtained from other sources" (Sentence 2) signals TEK's uniqueness; this is reinforced by the use of "cannot", an operator of epistemic modality relaying strong addresser commitment (for point of view and modality, see Simpson 1993, 43–47; summarised in Neary 2014, 181–182, and in Sect. 2.2.2). The two adjectives "broad" and "complex" (Sentence 4) convey its being wide-ranging and many-sided. The addresser argues that TEK "parallels the scientific discipline of Ecology" (Sentence 7) and "has been used to inform the field of Conservation Biology" (Sentence 8). Both the lexical verbs "parallels" and "inform" are highly positive here. The former does not merely indicate that TEK may slightly resemble ecology; it indicates that TEK matches or corresponds to ecology, which is termed a "scientific discipline" (see also "Western science" in the same sentence). The latter lexical verb does not simply denote that TEK may vaguely have inspired another academic area, conservation biology; it denotes that TEK permeates and is the essential principle of conservation biology.

Text 4 also comprises five instances of negative evaluation and modality in Sentence 6. After cataloguing TEK's favourable properties in Sentences 1–5, Sentence 6 treats, in negative terms, the attitude to it adopted by mainstream science. After opening with the adverb "Yet", implying contrast with the contents of the previous sentences, Sentence 6 depicts the involvement of Indigenous peoples in mainstream science as "rare" and contends that mainstream science "often overlooks and disparages" TEK. The adjective "rare" acquires a negative connotation in this context, whereas the two lexical verbs "overlooks" and "disparages" are negative in most, if not all, contexts. As shown in Sects. 3.2.1 and 4.2.1, in line with the Hallidayan view of modality (Halliday 2014, 176–183), the adjective "rare" and the adverb "often" indicate values of usuality and oftenness ("sometimes/usually/always"); as such, they are also embraced in the epistemic modal system and relay the addresser's psychological point of view. More precisely, "rare" suggests a low value of Indigenous peoples' involvement, while "often" evokes a medium value of mainstream science overlooking and disparaging TEK: this evidence serves to confirm the opposition between mainstream and TEK approaches to ecology. As a result, the addresser of Text 4 does not blame mainstream science for its attitude by employing more explicit stylistic devices, for example, such performative verbs as "accuse" or "reproach"; nevertheless, her criticism is brought to light by her lexical and modality choices in Sentence 6 and by all the linguistic data in favour of TEK she provides in the other sentences.

In addition, the addresser of Text 4 also indirectly criticises mainstream science through one of the distinctive elements of its language: the academic and disciplinary practice of citation (Hyland 2000, 22–29). The text contains four cases of citation. On the one hand, "(Gadgil et al. 1993)" (Sentence 4) and "(Westley and Miller 2003)" (Sentence 6) were defined by Hyland as non-integral citations, given that the authors' names appear in parentheses; this pattern is preferred by writers working in the hard disciplines, like the addresser of Text 4, who researches ecology. On the other hand, Hyland called "According to Huntington (1998)" (Sentence 2) and "According to many scholars" (Sentence 7) integral citations, as the citing clauses incorporate the names of the cited authors; to be more precise, these are non-subject integral citations, since the authors' names are quoted as adjunct agents; when the hard disciplines do not opt for non-integral citations, this is the pattern they prefer.

When citations are investigated from the disciplinary perspective of stylistics and point of view, they convey that the knowledge, beliefs and cognition of the addresser of Text 4 are based on the research on ecology undertaken by her colleagues. Consequently, citations can be scrutinised as operators of epistemic modality: they show that the addresser is committed to the veracity of the propositions she is communicating, because they are validated by her colleagues' findings. As well as "cannot" (Sentence 2), these linguistic indicators express high levels of epistemic modality and strong addresser commitment to her propositions; hence, they do not constitute negative modal shading which, by contrast, is realised by low levels of epistemic modality signalling the addresser's uncertainty (for modal shading, see Simpson 1993, 50–69; summarised in Neary 2014, 183–187, and in Sect. 2.2.2).

In Text 4, while modality figures in four sentences out of eight (Sentences 2, 4, 6 and 7), it does not appear in the remaining four sentences (Sentences 1, 3, 5 and 8). The latter are unmodalised categorical assertions: they constitute neutral modal shading and indicate the strongest possible value of addresser commitment. That is to say, the addresser of the text does not psychologically judge what she maintains about ecology and TEK. The stylistic role of these modality and shading patterns (no negative modal shading plus neutral modal shading) is to unequivocally suggest the truthfulness of the addresser's propositions founded on scientific data; accordingly, to convince her addressee that TEK must play an integral part in the discipline of ecology, and that Indigenous peoples' experience must be highly appreciated and broadly spread in the academic community.

The unmodalised categorial assertion in Sentence 5 includes the compressed noun phrase "time-tested resource management systems of long-resident peoples", which in turn includes the noun "resource". In the scientific context of Text 4, this noun refers to "natural resources" as delineated by the ecology textbook *Essentials of Ecology*, viz. "Materials such as air, water, and soil and energy in nature that are essential or useful to humans" (Miller and Spoolman 2019, G8; see also Harré et al. 1999, 40 and Sect. 3.1). From a stylistic viewpoint, the noun can be explored as a conceptual metaphor (Lakoff and Johnson 2003; see Sect. 2.2.3), more precisely as a conventionalised and lexicalised metaphor (Goatly and LLE Project 2002–2005). Indeed, it was recorded in MetaNet (David et al. 2016–2018), one of the three metaphor databases referenced in this book, under several metaphor entries, both as a target domain and as a source domain, and under several frame entries. The most pertinent to the ecological discourse of Text 4 are the following:

1. Metaphor NATURE IS A RESOURCE, illustrated by the example texts "we must deepen our understanding of the economic value of nature and invest in the vital services that it provides for us, if we are to have a truly sustainable economic future" and "To start with, everything we are made of and everything we make comes from the environment";
2. Metaphor CONSUMING RESOURCES IS EATING, exemplified by the sentences "This SUV sure is a gas guzzler" and "Demand for paper goods is eating away at our planet's forests";

3. Metaphor CONSUMING TOO MANY RESOURCES IS EATING TOO MUCH FOOD, an entailment of the previous metaphor, illustrated by the example texts "Developing nations are gobbling up our fossil fuels at any [sic] alarming rate" and "The United States' insatiable appetite for electronics will entirely consume our supply of rare metals";
4. Frame Natural Resource, which is a subcase of the related frame Resource (for frames, see Tannen 1993. In accordance with common typographical conventions, the initials of frames are Capitalised, and metaphors, targets and sources are in SMALL CAPITALS).

The frame Natural Resource confirms that the noun "resource" in Text 4 denotes "natural resources" as defined by *Essentials of Ecology*, namely natural materials vital to us humans. In the conceptual metaphor NATURE IS A RESOURCE, therefore in its example texts, the natural world is portrayed as a stock or reserve of animals, plants and materials which, when necessary, can be drawn on by us humans. The conceptual metaphor CONSUMING RESOURCES IS EATING, together with its entailment CONSUMING TOO MANY RESOURCES IS EATING TOO MUCH FOOD and their example texts, represents the use (and abuse) of these animals, plants and materials as the consumption (and overconsumption) of food and drink. Underpinning these two metaphors is another conceptual metaphor, RESOURCES ARE FOOD AND DRINK. Here, two key characteristics of the source domain FOOD AND DRINK map to the target domain RESOURCES: they are elements indispensable and fundamental to human life and more-than-human life alike, and they must be consumed abstemiously and sensibly to avoid health problems (metaphorically, ecological issues). As a result, the compressed noun phrase "time-tested resource management systems of long-resident peoples" from Text 4 conveys that Indigenous peoples have learnt all the strategies and procedures to utilise the natural materials in their physical environments in a fruitful and sustainable way. Given that these materials are as vital as food and drink to all human, animal and vegetable species, the academic field of ecology must, in turn, learn from Indigenous peoples and their experience which has stood the test of time.

6.2.2 Summary and Evaluation

Text 4, drawn from National Park Service's website and containing the marker word "ecology", and the stylistic techniques of foregrounding, point of view and metaphor featuring in the text will be summarised and assessed in this section, following the ecostylistic analysis of the text and its stylistic mechanisms in the previous section. To be more precise, the value-system of Text 4 brought to light by this analysis is appraised here against the ecosophy of this research and its norms; consequently, it is shown whether the text and the devices deployed by the addresser are, in fact, beneficial.

As noted in Sect. 6.2.1, underlying Text 4 and its idea of ecology is the concept of Traditional Ecological Knowledge, or TEK, which is employed to delineate the marker word "ecology" and the scientific methods characterising this area (Sentence 7). The concept of TEK and the experience of Indigenous peoples as described and celebrated in the text are beneficial and follow the ecosophy. Valuing and reinforcing local societies, traits and habits recalls norm no. 6: Social Justice, while Indigenous practices can serve as models to develop new economic and social systems based on ecological principles and fair and equitable goals, thereby aligning with norm no. 7: Adaptation.

But what is the lexicological and naturalistic definition of the marker word "ecology", hence of the adjective "ecological" incorporated in the acronym "TEK"? The beneficial definition offered by sense 1.a in the Oxford English Dictionary (OED Online 2022) asserts that the academic discipline of ecology explores the interactions of living organisms with their surroundings; the noun also denotes the interactions themselves, particularly those of a specific organism. Furthermore, *Ecology Basics* (Editors of Salem Press 2004, 171, 176) claims that the thrust of ecology is to study scientific laws and to adopt them in our daily lives and in our ecological spaces. Text 4 and its ecostylistic examination have demonstrated that TEK is already acquainted with a large number of ecological principles, and that Indigenous peoples invariably adhere to them in their environments. Accordingly, this empirical field and the lifestyle of these peoples are beneficial and in accordance with the ecosophy and its norms no. 1: Valuing Living and no. 5: Environmental

Limits. These beneficial notions should be popularised among the public at large, along with the stylistic techniques relaying them in Text 4: foregrounding, point of view and metaphor frequently co-occur to highlight major words, phrases and clauses and to make them more prominent.

As demonstrated in Sect. 6.2.1, in Text 4 foregrounding and overspecification are realised by four noun phrases showing articulate premodification and postmodification, or compressed noun phrases, like "ecological information and insights relevant to ecological management and research that cannot be obtained from other sources" (Sentence 2). They typify academic and scientific writing, and are utilised to communicate a large number of details, here about TEK, in a small number of words. The fact that these noun phrases are compressed increases the time and effort expended by the addressee to read the information in the phrases. This is not a beneficial stylistic mechanism per se; it becomes so when, with the authoritative and reliable academic language of ecology, it expresses that TEK applies to mainstream ecological research and management (Sentence 2) and supplies ecological skills and knowledge other sources are unable to supply (Sentence 2); and when it signals that these skills and knowledge are vast, multi-layered and unrivalled in their being unique to specific environments and in their being proven over a long period of time (Sentences 4 and 5). All these messages agree with the ecosophy and its norms no. 6: Social Justice and no. 7: Adaptation. Local civilisations, aspects and procedures must be considered favourably and reinforced; this should also concur in reducing the phenomenon of slow violence (Nixon 2011) against Indigenous peoples; moreover, these civilisations, aspects and procedures must be regarded as exemplars when we humans from the Global North adapt to unpromising ecological circumstances, and when we strengthen our response to environmental change in the near future.

In Text 4, a similar function to compressed noun phrases is carried out by the foregrounding strategy of the repetition of syntactic structures in four two-part coordinated structures with the same or corresponding meanings, like the non-finite clauses "to sustain themselves and to maintain their cultural identity" (Sentence 3). Indeed, to the advantage of the addressee, who may not be conversant with ecology or TEK, repetitions gloss and emphasise beneficial ideas in the text. For instance, these syntactic structures underline the value of Indigenous cultural identity (Sentence 3), the complexity of the Indigenous ecological systems which have elaborated TEK (Sentence 6), empirical TEK sharing its methods with the scientific field of ecology (Sentence 7), and mainstream science often disregarding and deprecating TEK (Sentence 6). This additional information, together with the message of the entire Text 4, indicates that TEK must not be disregarded or deprecated, and that this information and the text are in line with two more norms in the ecosophy: norms no. 1: Valuing Living and no. 2: Wellbeing. It can be inferred that the ecosphere as a whole is taken into account in Indigenous ecological systems and in their TEK, as it is by ecology; in addition, if Indigenous peoples have been able to evolve and maintain an identifiable cultural identity, it can also be inferred that TEK allows them not only to exist and survive, but also to enjoy physical and mental prosperity.

The addresser of Text 4 also deploys the stylistic device of point of view, which often interacts with foregrounding. In Sentence 8, four indicators of place, time and person deixis create empathy between the addresser and her addressee, and explicitly involve both in the text. Therefore, this inclines the addressee to also share the addresser's ideological point of view on TEK and her tribute to it, which are suggested by means of positive evaluative words and phrases. Several of them were studied above as constituents in compressed noun phrases and in two-part coordinated syntactic structures, for example, the relative clause "that cannot be obtained from other sources" (Sentence 2) and the two adjectives "broad" and "complex" (Sentence 4). Furthermore, the two positively value-laden lexical verbs "parallels" (Sentence 7) and "inform" (Sentence 8) confirm TEK's scientific methods and approaches in its being equivalent to ecology and in its permeating conservation biology. The accuracy of these methods and approaches is conveyed by the positive evaluative prepositional phrases "For thousands of years" (Sentence 3) and "around the world" (Sentence 4), and adjective phrase "time-tested" (Sentence 5). This evidence recalls the ecosophy and its norm no. 3: Now and the Future. Positively value-laden words and phrases praise TEK, since it has turned out to be successful in time and place; as a result, it is most likely to be equally effective in the future, and must be employed without delay as an ecological model to save the earth from the current environmental crisis.

Ideological and psychological point of view are utilised by the addresser of Text 4 to criticise mainstream science for not accepting TEK; the former type of point of view is relayed by three negative evaluative words, the latter by two epistemic modal operators and four instances of citation. As shown in Sect. 6.2.1, these words and operators are the adverb "Yet", the epistemic modal adjective "rare", the epistemic modal adverb "often", and the two lexical verbs "overlooks" and "disparages" (all in Sentence 6); the citations are non-integral "(Gadgil et al. 1993)" (Sentence 4) and "(Westley and Miller 2003)" (Sentence 6), and non-subject integral "According to Huntington (1998)" (Sentence 2) and "According to many scholars" (Sentence 7). The addresser's accusation against mainstream science is indirect but emerges clearly from these expressions as deployed in the context of Text 4. The tough stance on TEK adopted by mainstream science is destructive when compared with the ecosophy, as rejecting TEK entails rejecting the ecosophy and its various norms followed by TEK; consequently, the addresser's reproach is a beneficial stylistic practice.

As noted in Sect. 6.2.1, the linguistic indicators of epistemic modality and the examples of citation in Text 4 do not constitute negative modal shading; moreover, the unmodalised categorical assertions in four sentences out of eight (Sentences 1, 3, 5 and 8) realise neutral modal shading. These modality and shading patterns contribute to constructing the addresser's disciplinary identity as a researcher on ecology: she is professional and authoritative, because what she observes about TEK and Indigenous peoples is either referenced or presented as factual. This reinforces her criticism of mainstream science and constitutes a beneficial stylistic technique, given that it gives prominence to the ecosophy and to the norms TEK aligns with.

In Sentence 5, the stylistic mechanism of metaphor co-occurs with foregrounding and point of view; more precisely, the conventionalised and lexicalised metaphor "(natural) resource(s)" can be found in the compressed noun phrase "time-tested resource management systems of long-resident peoples" (Sentence 5), which is a constituent of an unmodalised categorial assertion. This is an extremely recurrent metaphor in general discourse and ecological discourse alike; for instance, it also occurs three times in the beneficial Text 2 investigated in this book (see Sect. 4.2). As shown in Sect. 3.1, in the academic discourse of ecology and environmental studies, the noun phrase "natural resources" indicates natural materials (e.g. air, water, soil, energy) which are fundamental or helpful to us humans. This is a destructive concept against the ecosophy and its norms no. 1: Valuing Living and no. 4: Care. Firstly, it communicates the anthropocentric mindset that the purpose of the components of the ecosphere is to sustain our human lives only, not animal or vegetable lives as well; it resists the ecocentric mindset that these components are an end in themselves. Secondly, the concept of natural resources expresses that these components are exclusively entities instrumental to our human lives and wellbeing; it counters the notion that they have inherent value or value for other organisms.

Hence, as noted in the ecosophy and in its norm no. 1: Valuing Living, the destructive metaphorical term "(natural) resource(s)" must be replaced with the value-neutral non-metaphorical terms "(natural) materials" and "(natural) processes" or, merely, "nature" and "the natural world". In other words, the language system must be altered: a natural material must not be depicted via the MetaNet destructive frame Natural Resource, nor must nature be portrayed via the MetaNet destructive conceptual metaphor NATURE IS A RESOURCE any longer. If the language system is altered, it will not be able to produce such destructive sentences as the first MetaNet example text for the NATURE IS A RESOURCE metaphor, i.e. "we must deepen our understanding of the economic value of nature and invest in the vital services that it provides for us, if we are to have a truly sustainable economic future".

In addition, if the language system is altered, destructive metaphors founded on "(natural) resource(s)" (e.g. CONSUMING RESOURCES IS EATING, CONSUMING TOO MANY RESOURCES IS EATING TOO MUCH FOOD, RESOURCES ARE FOOD AND DRINK) will turn into beneficial metaphors founded on "natural materials": CONSUMING NATURAL MATERIALS IS EATING, CONSUMING TOO MANY NATURAL MATERIALS IS EATING TOO MUCH FOOD, NATURAL MATERIALS ARE FOOD AND

DRINK. These beneficial metaphors lose their anthropocentric qualities, but retain their ecocentric qualities. The FOOD AND DRINK source domain properties now map to the new NATURAL MATERIALS target domain, and signal that these materials are essential to all humans, animals and plants; as such, as prescribed by the ecosophy and its norms no. 1: Valuing Living and no. 4: Care, they must be employed sparingly and wisely. As disclosed by scrutinising Text 4 and its stylistic strategies from an ecostylistic standpoint, and by comparing its message with the ecosophy of this research, Indigenous peoples, with the TEK and ways of life they have developed, have long been familiar with these beneficial metaphors; accordingly, the scientific area of ecology must more extensively rely on TEK and on these metaphors too, as put forward by National Park Service's environmental website.

References

Biber, D., and B. Gray. 2010. Challenging stereotypes about academic writing: Complexity, elaboration, explicitness. *Journal of English for Academic Purposes* 9 (1): 2–20.

Biber, D., S. Johansson, G. Leech, S. Conrad, and E. Finegan. 2021. *Grammar of Spoken and Written English*. Amsterdam and Philadelphia: John Benjamins.

David, O., G. Lakoff et al. 2016–2018. *MetaNet*. Berkeley: University of California. https://metaphor.icsi.berkeley.edu/pub/en. Last accessed April 2022.

Editors of Salem Press, ed. 2004. *Ecology Basics*. Pasadena, CA, and Hackensack, NJ: Salem Press.

Emmott, C., and M. Alexander. 2016. Defamiliarization and foregrounding: Representing experiences of change of state and perception in neurological illness autobiographies. In *The Bloomsbury Companion to Stylistics*, ed. V. Sotirova, 289–307. London: Bloomsbury.

Gadgil, M., F. Berkes, and C. Folke. 1993. Indigenous knowledge for biodiversity conservation. *Ambio* 22: 151–156.

Goatly, A., and LLE Project. 2002–2005. *Metalude*. Hong Kong: Lingnan University. http://www.ln.edu.hk/lle/cwd/project01/web/home.html. Last accessed April 2022.

Halliday, M.A.K. 2014 [1985]. *Halliday's Introduction to Functional Grammar*, revised by C.M.I.M. Matthiessen, 4th ed. London and New York: Routledge.

Harré, R., J. Brockmeier, and P. Mühlhäuser. 1999. *Greenspeak: A Study of Environmental Discourse*. London: Sage.

Huntington, H.P. 1998. Observations on the utility of the semi-directive interview for documenting TEK. *Arctic* 51: 237–242.

Hyland, K. 2000. *Disciplinary Discourses: Social Interactions in Academic Writing*. London: Longman.

Kemp, D.D. 2002 [1998]. *The Environment Dictionary*, 2nd ed. London and New York: Routledge [electronic edition].

Lakoff, G., and M. Johnson. 2003 [1980]. *Metaphors We Live By: With a New Afterword*, 2nd ed. Chicago: University of Chicago Press.

Leech, G., and M. Short. 2007 [1981]. *Style in Fiction: A Linguistic Introduction to English Fictional Prose*, 2nd ed. Harlow: Longman.

Miller, G.T., and S.E. Spoolman. 2019 [2007]. *Essentials of Ecology*, 8th ed. Belmont, CA: Cengage.

Mukařovský, J. 1964. Standard language and poetic language. In *A Prague School Reader on Esthetics, Literary Structure and Style*, ed. P.L. Garvin, 17–30. Washington, DC: Georgetown University Press.

Mukařovský, J. 1977. On poetic language. In *The Word and Verbal Art: Selected Essays*, ed. J. Burbank and P. Steiner, 1–64. New Haven, CT and London: Yale University Press.

Myerson, G., and Y. Rydin. 1996. *The Language of Environment: A New Rhetoric*. London and New York: Routledge.

Neary, C. 2014. Stylistics, point of view and modality. In *The Routledge Handbook of Stylistics*, ed. M. Burke, 175–190. London and New York: Routledge.

Nixon, R. 2011. *Slow Violence and the Environmentalism of the Poor*. Cambridge, MA: Harvard University Press.

Oxford English Dictionary (OED) Online. 2022. Oxford: Oxford University Press. https://www.oed.com/. Last accessed April 2022.

Short, M. 1996. *Exploring the Language of Poems, Plays and Prose*. Harlow: Longman.

Simpson, P. 1993. *Language, Ideology and Point of View*. London and New York: Routledge.

Simpson, P. 2010. Point of view. In *Language and Style: In Honour of Mick Short*, ed. D. McIntyre and B. Busse, 293–310. Basingstoke: Palgrave MacMillan.
Smith, T.M., and R.L. Smith. 2015. *Elements of Ecology*, 9th ed. Harlow: Pearson.
Stibbe, A. 2021 [2015]. *Ecolinguistics: Language, Ecology and the Stories We Live By*, 2nd ed. London and New York: Routledge.
Tannen, D., ed. 1993. *Framing in Discourse*. Oxford: Oxford University Press.
Westley, F.R., and P.S. Miller, eds. 2003. *Experiments in Consilience: Integrating Social and Scientific Responses to Save Endangered Species*. Washington, DC: Island Press.

7
Sustainability

7.1 Definition and Evaluation

In this section, I explore the meaning of "sustainability", the last of the five marker words (Myerson and Rydin 1996, 6; see Sect. 1.1) examined in this book. The noun "sustainability" was preferred to the noun phrase "sustainable development": when applied to nature and the environment, "sustainable development" is at best ambiguous, at worst contradictory (Myerson and Rydin 1996, 97–107). In sense 3.a of the Oxford English Dictionary (OED Online 2022), the adjective "sustainable" is defined as "Capable of being maintained or continued at a certain rate or level". As anticipated in Sect. 2.1.3, the noun "development" commonly refers to economic growth, industrialism and capitalism and reflects the explicit bias towards them in modern Western culture. Therefore, not only is economic growth not identified as a problem, but it is recognised as the entity to be sustained, namely "maintained or continued". In fact, the notion of sustainable development is ecologically flawed, for it focuses on "development" rather than "environment" or "society" (for criticism of the metaphorical three-legged-stool model of sustainable development,

D. F. Virdis, *Ecological Stylistics*,
https://doi.org/10.1007/978-3-031-10658-3_7

with the three legs being the environment, the economy and society, see Dawe and Ryan 2003).

Conversely, the noun "sustainability" can be applied to the "maintenance or continuation" of the natural systems and species supporting us humans; these systems and species must not be harmed or destroyed by our human actions and activities, but their viability and subsistence must be allowed to endure. As a result, preferring the noun "sustainability" to the ecologically inconsistent noun phrase "sustainable development" backgrounds economic and industrial interests and foregrounds environmental priorities, as is shown below and in the working definition at the end of this section (for sustainability as a figurative concept, see Larson 2011, 99–100; for the "newness" of the concept and practice of sustainability and sustainable development, see Myerson and Rydin 1996, 97–103).

Of the senses of the noun "sustainability" appearing in the Oxford English Dictionary, the following is the most appropriate to the matters considered in this book:

> **2.b.** *spec.* The property of being environmentally sustainable; the degree to which a process or enterprise is able to be maintained or continued while avoiding the long-term depletion of natural resources. (OED Online 2022)

The subordinating conjunction "while" connects two actions happening simultaneously, in this case maintaining or continuing a process or enterprise and "avoiding the long-term depletion of natural resources" (for the destructive noun phrase "natural resources", see Sect. 3.1; for an ecostylistic study of the noun "resource", see Sect. 6.2). Connecting the two actions means that certain processes and enterprises are likely to spark off the exhaustion of natural materials. Nevertheless, this part of the sense of the noun "sustainability" is lacking in detail and is, accordingly, vague and general. To be more precise, the text does not specify what particular processes and enterprises are alluded to here and does not argue that they are ordinarily human-initiated or human-induced.

Another part of the sense of "sustainability", however, is more informative and provides the detail that the only draining of natural materials to be prevented is that taking place in the long run. Underlying this is the first definition of "sustainable development", supplied in 1987 by the Brundtland Commission (see below): "Humanity has the ability to make development sustainable to ensure that it meets the needs of the present without compromising the ability of future generations to meet their own needs" (World Commission on Environment and Development [WCED] 1987, 16). Both "avoiding the long-term depletion of natural resources" and "the ability of future generations to meet their own needs" are problematic claims, as they express the question in the future trends: they do not push us humans into immediate and drastic action now and conceal the fact that ecological challenges are challenges at present, not only in the future. Semantically, the concept of "long-term depletion", in an instance of scalar implicature, logically entails the concepts of "short-term depletion" and "current depletion"; in other words, long-term depletion is the result of short-term depletion and current depletion. Therefore, the text should have asserted more plainly that *all* depletion of nature must be avoided.

The first part of the definition ("The property of being environmentally sustainable") is based on the sense of the related adjective "sustainable" in the Oxford English Dictionary:

> **2.b.** Designating forms of human activity (esp. of an economic nature) in which environmental degradation is minimized, esp. by avoiding the long-term depletion of natural resources; of or relating to activity of this type. Also: designating a natural resource which is exploited in such a way as to avoid its long-term depletion. Cf. SUSTAINABILITY *n.* 2b. (OED Online 2022)

This definition of the adjective "sustainable" explicitly ascribes to a human agent the agentless processes and enterprises vaguely referred to in the definition of the noun "sustainability": they are actually said to be "forms of human activity". These are mostly of an economic kind, namely they turn natural materials to profit. An economic connotation is also present in the lexical verb "exploited", in the last sentence of the definition, which designates the action of making use of natural materials and processes and utilising them for our human ends. Furthermore, the non-finite clause "avoiding the long-term depletion of natural resources", which can be found in both the definitions of the noun and of the adjective, in the latter is introduced by the preposition "by", indicating means: as a result, the clause denotes a strategy to accomplish the objective of minimising the phenomenon of ecological exhaustion. The lexical verb "minimized" implies that this phenomenon can be reduced to the smallest possible extent or degree, but cannot be eradicated definitively.

When evaluated against the ecosophy of this research and its norms (see Sect. 1.3), the notions of "sustainability" and "sustainable" as set out in the Oxford English Dictionary are controversial and, consequently, ambivalent (see Sect. 2.1.3); that is to say, they cover beneficial aspects (see Sect. 2.1.2) along with destructive aspects (see Sect. 2.1.4). The ecosophy, in its norm no. 4: Care, notes that we humans must be empathic to the other animal and vegetable organisms on the planet, and conscious that we have a deep impact on their lives and wellbeing; hence we must take responsibility for these circumstances. Accordingly, a part of the definition of the noun "sustainability" is made destructive by the fact that it does not acknowledge that we humans are the agents of the processes and enterprises it refers to; conversely, the part of the definition of the adjective "sustainable" admitting our human agency is beneficial.

As mentioned above, these human-initiated processes and enterprises in the definition of the noun are reworded as "forms of human activity (esp. of an economic nature)" in that of the adjective. The idea of the "economic nature" of our human actions and, in the next sentence, that of their exploitation of natural materials communicate that they capitalise on the natural world. The ecosophy and its norms no. 1: Valuing Living and no. 4: Care observe that animals and plants have inherent value and must not be thought of as resources to turn to our human account. Here the fact that the two definitions, mainly that of "sustainable", are ambivalent emerges distinctly. On the one hand, acknowledging our human economic interests and advantage, i.e. our radical effect on the earth, is a beneficial attribute of the adjective. On the other hand, playing down such human economic interests and advantage by referring to the avoidance of long-term depletion and minimised environmental degradation is a destructive attribute of the adjective. This is because, given that nature has worth in itself, it must ideally be conserved for the time being, in the short run and in the long run, and must not be damaged at all.

Realistically, however, this purpose of safeguarding the more-than-human world cannot be expected to be attained without difficulty. The concepts of avoiding long-term depletion and minimising environmental degradation as phrased in the two definitions can be scrutinised through at least four of the norms in the ecosophy. On the one hand, norm no. 3: Now and the Future states that the ecosophy does not have any time limitations, primarily with regard to the wellbeing of human, animal and vegetable species. Moreover, in norm no. 5: Environmental Limits, it is argued that, if the contents of the ecosophy are not applied, our conduct will shortly hinder the natural processes of the more-than-human world, including its capacity to replenish itself. On the other hand, norm no. 2: Wellbeing points out that the fulfilment of the necessities of our human life is one of the core and concrete prerequisites to the welfare of us humans and the other animal and vegetable organisms alike. In addition, Norm no. 4: Care incorporates the value of regret, which entails cutting down our impact on the planet and the harm we do to it. Therefore, considering these norms, avoiding the utter depletion of the natural world seems to be an extremely ambitious but achievable aim. Nevertheless, even two of the norms in the ecosophy (no. 2 and no. 4) have to practically admit that all environmental degradation cannot be conclusively eradicated, but can only be played down and curbed, for our deleterious influence on nature is apparently inevitable.

A definition of the noun "sustainability" features in nearly all the recently published or reprinted ecology and environment dictionaries and textbooks I referred to. Three of them (Allaby 2015; Editors of Salem Press 2004; Kemp 2002) inform their readers about the development and success of the notion of sustainability. More precisely, the first maintains that

> The concept was introduced in the late 1970s and was emphasized strongly in the *World Conservation Strategy*, published in 1980 by the

> IUCN [International Union for Conservation of Nature] in collaboration with the UN Environment Programme and the World Wildlife Fund (now the World Wide Fund for Nature). *Our Common Future*, published in 1983 [sic] by the World Commission on Environment and Development (the Brundtland Commission), defined it as development that "seeks to meet the needs and aspirations of the present without compromising the ability to meet those of the future". (Allaby 2015, 395)

The second text adds that "Since publication of the Brundtland Commission's report, sustainable development has become the dominant global position on the environment, ecology, and economic development" (Editors of Salem Press 2004, 618).

Most of the definitions of "sustainability" in the dictionaries and textbooks resemble those in the Oxford English Dictionary in length. With regard to their contents, they are also ambivalent. On the one hand, it may be beneficial to acknowledge that we humans have peculiar cultural systems, economies and needs; to take account of the environmental consequences of these economic activities; to assert that changes in our attitude and behaviour are necessary; and to mention that we should favour natural materials which can be replaced or renewed in order not to deplete them. On the other hand, these potentially beneficial elements are made destructive when the texts do not explicitly criticise the fact that economic requirements are portrayed as being as important as ecological requirements; that the only needs of the world's population to be borne in mind are consumption needs; that our human activities cause harm to the environment; that our cultural systems and economies change the earth's environmental conditions; that our human current consumption needs and economic production in the industrialised and overdeveloped countries are already galloping beyond environmental limits and must be decreased; that the changes in our attitude and behaviour must be more impactful than employing environment-friendly technology; and that natural materials, both renewable and non-renewable, must be treated thoughtfully and with proper attention, as they have intrinsic value (see, e.g. Allaby 2015, 395; Bailey 2004, 224; Editors of Salem Press 2004, 618; Kemp 2002, 389; Miller and Spoolman 2019, G12).

One of the ecology textbooks offers further information than the other texts about the idea of sustainability. To be more precise, the chapter "Sustainable development" in *Ecology Basics* (Editors of Salem Press 2004, 618–620) makes a distinction between "weak sustainability" and "strong sustainability" (for the weak-strong sustainably spectrum and the economic perspectives on it, see Myerson and Rydin 1996, 103–107). The former is defined as follows:

> Weak sustainability requires that depletions in natural capital be compensated for by increases in human-made capital of equal value. For example, the requirements for weak sustainability are met when a tree (natural capital) is cut for the construction of a frame house (human-made capital). (Editors of Salem Press 2004, 618)

At the core of the definition is the concept of compensation, viz. counterbalancing or making amends for damage; in this case, the depletion of nature must be compensated for by a rise in manufactured objects (for the destructive noun phrase "natural capital", see Sect. 3.1). The noun phrase "depletions in natural capital" triggers the existential presupposition that the draining of natural materials exists. This presupposition is also a part of the ecosophy of this research. As previously shown, its norms no. 2: Wellbeing and no. 4: Care admit that this draining exists, can be minimised but cannot be fully prevented. What is not a part of the ecosophy, however, is the idea that cutting a tree, namely a natural living being, is of "equal value" to and can be counterbalanced by constructing a house, namely a lifeless artificial object. A frame house is certainly more eco-friendly than a building of human-made non-renewable materials; nevertheless, even more certainly, a tree and a house (albeit a frame one) do not play equivalent ecological parts in the environment surrounding them and in the entire ecosystem comprising them. As a result, the definition of the noun phrase "weak sustainability" is not in line with the ecosophy; it is, consequently, destructive, so much so that it is difficult to reword it in a beneficial way. In short, "weak sustainability" is not sustainability at all for the ecosophy.

A partial rewording of the definition of "weak sustainability" is supplied by the definition of "strong sustainability" in the same textbook:

> Strong sustainability requires that depletions of one sort of natural capital be compensated for by increases in the same or similar natural capital. For example, the requirements for strong sustainability are met when a tree is cut and a new tree is planted to replace it, or when loss of acreage in equatorial rain forests in Brazil is compensated for by an increase in the acreage of temperate rain forests on the Pacific coast of North America. (Editors of Salem Press 2004, 618)

Like the notion of weak sustainability, underpinning that of strong sustainability is the idea of compensation for the depletion of natural materials and the existential presupposition that this depletion exists; as mentioned above, this presupposition is also shared by the ecosophy. The definition offers two examples of what it terms "strong sustainability"; the first consists of planting a tree to replace another which was felled. The text contains the two noun phrases "a tree" and "a new tree", which are expressions referring to the two plants. These referring expressions are utilised attributively, meaning "whatever tree fits the description"; in fact, the quantifier "any" could be substituted for the indefinite article "a" in both phrases. The first example hence implies that any tree of a given species can be replaced by any other tree of the same species, but it also implies that a tree of any species can be replaced by a tree of any other species. In other words, this logically entails that the requirements for strong sustainability are also met when, for instance, a tree typical of the Indonesian rainforest is felled and an oil palm substituted for it. The only explanation for this procedure is not ecological, i.e. preserving the rainforest, but economic, i.e. ruining the rainforest to use the newly-planted oil palms to our human advantage. The second example of strong sustainability is constituted by growing the extent of a type of forest (here temperate rain forests on the North American Pacific coast) to counterbalance the declining extent of a dissimilar type of forest (here Brazilian equatorial rain forests). Like that in the first example, this process can only be carried out on economic grounds, not on ecological grounds: its principal goal is a land-clearing project concerning equatorial rain forests to exploit these lands, not a reforestation project regarding temperate rain forests to protect them.

Taking the ecosophy of this research into consideration, its norm no. 1: Valuing Living notes that the ecosophy embraces all humans, animals and plants and makes no differentiation between them, irrespective of whether the nonhuman species can be deployed by our human species or not. It is observed in norm no. 2: Wellbeing that flourishing and welfare must be high for us humans. However, norm no. 4. Care points out that we must feel regret at the animal and vegetable organisms we use to obtain our wellbeing and, therefore, cut down the numbers of those we kill or harm; the same norm also states that we must feel gratitude to and recompense them somehow. Accordingly, when assessed against the ecosophy, the definition of "strong sustainability" may seem to be beneficial at first reading, since a tree is compensated for by another tree and not by a frame house. Nevertheless, even at first reading it comes out that the ecological function of a tree in Brazil does not correspond to that of a tree on the North American Pacific coast, because each of them is a part of a distinct environment and ecosystem. At second reading, the implications of the definition emerge, running counter to at least norms no. 1: Valuing Living and no. 4: Care and their concepts of equality, regret and gratitude; what is more, these implications can be employed to disguise and greenwash financial priorities.

As a result, a beneficial rewriting of the definition of "strong sustainability" must, firstly, leave out the adjective "strong" and the distinction from the notion of weak sustainability, given that, as previously shown, this notion is not sustainable according to the ecosophy. Secondly, this beneficial rewriting must cancel the destructive contents and implications in the definition. Thirdly, for the sake of clarity and completeness, it must be introduced and complemented by the definition of the adjective "sustainable" in the Oxford English Dictionary quoted and scrutinised above. Consequently, the beneficial rewriting must argue something like the following:

> Sustainability designates forms of human activity (esp. of an economic nature) in which environmental degradation is minimised, esp. by avoiding the depletion of natural materials. Sustainability also designates the use of a natural material in such a way as to avoid its depletion.

> Sustainability requires that depletions of one sort of natural material be compensated for by increases in the same natural material or in a very similar natural material with the same ecological function in the ecosystem, so that the ecological balance of the ecosystem is kept and the ecosystem itself is conserved now and in the future. For example, the requirements for sustainability are met when a tree is cut and replaced by a new tree of the same species or of a species belonging to the same ecosystem as the felled tree; the new tree must be planted in the same or nearby environment as the felled tree.

It is this working definition of the noun "sustainability", which is more in accordance with the ecosophy of this research, which will be taken as a starting point for the ecostylistic scrutiny in the section below.

7.2 "Sustainability" in Navdanya's Words

7.2.1 Ecostylistic Analysis

The marker word "sustainability" has now been defined from the lexicological standpoint of the Oxford English Dictionary (OED Online 2022) and from the naturalistic standpoint of ecology and environmental studies. How is this marker word deployed in the environet (Myerson and Rydin 1996, 7; see Sect. 1.1), particularly in the environmental website and discourse of Navdanya International (https://navdanyainternational.org/) (see Sect. 1.2), and what beneficial ecostylistic mechanisms are connected with it? Text 5 is taken from the article "A simplicity revolution: An ecological and ethical imperative to protect life on earth and ensure justice for all beings". The article was written by Navdanya's founder Vandana Shiva and published on the movement's website on 11 July 2020, in celebration of World Simplicity Day on the following day. The text consists of four paragraphs from the article's opening section, more precisely its third to sixth paragraphs.

Text 5

[1] The Corona lockdown has forced us to shed the clutter and the speed and become an invitation to humanity to differentiate between the essential and non-essential.

[2] In making us stay home, the crisis creates the possibility of realising that no matter who we are, where we are, we are members of one Earth Family and share a common home, our beautiful Earth. [3] We are part of Nature, not separate from it nor superior to it.

[4] Living on the earth with other beings makes simplicity (shedding the clutter) an ethical and ecological obligation. [5] By taking simply what we need to sustain our life, living within ecological and planetary boundaries, we leave vital ecological space for other beings.

[6] All beings need their share of ecological space and the right to participate in the living processes which ensure food and water for all. [7] By extracting more from the earth, we overstep the limits of our rightful share, disrupt planetary boundaries, ecological limits as well as the integrity of species. [8] In the web of life species sustain each other. [9] Sustainability and justice are interconnected, and robbing others of their rightful share deprives them of their basic needs with a deepening crises [sic] of food and water, poverty, hunger and starvation.

(Copyright © 2022 Navdanya International. "A simplicity revolution: An ecological and ethical imperative to protect life on earth and ensure justice for all beings". By Dr Vandana Shiva. Posted 11/07/2020 at 10:02 am. Available at https://navdanyainternational.org/world-simplicity-day-12th-july-2020/. Last accessed April 2022)

[205 words]

The marker word "sustainability" occurs once in Text 5, in the concluding Sentence 9: "Sustainability and justice are interconnected, and robbing others of their rightful share deprives them of their basic needs with a deepening crises [sic] of food and water, poverty, hunger and starvation". Its root, the lexical verb "sustain", is present twice in two different sentences: "By taking simply what we need to sustain our life [...]" (Sentence 5), and "In the web of life species sustain each other" (Sentence 8). Of the definitions of the marker word "sustainability" included in the Oxford English Dictionary and in the ecology and environment dictionaries and textbooks analysed in Sect. 7.1, which is or are the most relevant to Navdanya's value-system and to the ecological discourse of Text 5? As shown in that section, all the definitions of the noun "sustainability" examined there, as well as that of the adjective "sustainable" in the Oxford English Dictionary, are disputed and, therefore, ambivalent when compared with the ecosophy of this research and its norms (see Sect. 1.3), especially norms no. 1: Valuing Living, no. 3: Now and the Future, no. 4: Care and no. 5: Environmental Limits. Furthermore, the definition of "weak sustainability" provided by *Ecology Basics* (Editors of Salem Press 2004, 618) and even that of "strong sustainability" in the same textbook (Editors of Salem Press 2004, 618) were evaluated as downright destructive. For these reasons, Sect. 7.1 proposed a working definition aligning with the ecosophy: this definition rephrases that of "strong sustainability" in a beneficial way and completes it with the beneficial ideas in the Oxford English Dictionary definition of the adjective "sustainable".

Another definition of the noun "sustainability" in accordance with the ecosophy can be inferred from Text 5, particularly from the three sentences with the instance of this noun and the two instances of the lexical verb "sustain". The main clause in Sentence 5 states that we humans must "leave vital ecological space for other beings"; Sentence 8 clarifies that "In the web of life species sustain each other"; Sentence 9 concludes this line of reasoning by deducing that "Sustainability and justice are interconnected […]". As a result, a beneficial definition of "sustainability" based on Text 5 can be the following:

> Sustainability designates forms of human activity in which we humans take from the earth simply the materials we need to sustain our life, which allow us to live within ecological and planetary boundaries, and which, consequently, leave vital ecological space for other human, animal and vegetable beings. Underpinning the concept of sustainability is the premiss that in the web of life species sustain each other. It follows that sustainability and justice are interconnected, since robbing others of their rightful share deprives them of their basic needs with a deepening crisis of food and water, poverty, hunger and starvation.

This definition agrees with the ecosophy and with four of its norms. It takes into account the lives and flourishing of all human, animal and vegetable beings, namely the entire ecosphere (norms no. 1: Valuing

Living and no. 2: Wellbeing); it argues that we humans must only take from the planet what we require to live, thereby prioritising quality over quantity and physical and psychological health over economic expansion (norm no. 2: Wellbeing); it states that all species sustain one another in the ecosystem, which is an essential condition for prosperity (norm no. 2: Wellbeing); it considers ecological and planetary boundaries and alludes to the fact that we humans squander goods, materials and energy and produce waste (norm no. 5: Environmental Limits); and it contends that, and explains why, sustainability and justice are inextricably linked, hence it suggests that we lessen inequality and share global wealth more equally (norm no. 6: Social Justice). These and further beneficial notions are conveyed in Text 5 via a number of stylistic strategies, i.e. foregrounding, point of view and metaphor, introduced in the theoretical Sect. 2.2.

The first foregrounding device in Text 5 can be found in the opening Sentence 1, in the non-finite clause "to shed the clutter and the speed". At first reading, "shed the clutter" seems to be a standard collocation with the meaning of "get rid of disorder", and the noun "clutter" seems to have the literal meaning of "litter, jumbled objects". However, "clutter" is coordinated with "speed", an abstract noun which does not standardly collocate with the lexical verb "shed" and which, in this ecological context, refers to our human habit of getting things done in great haste and without proper care. This leads the addressee of Text 5 to understand that "clutter" is also deployed figuratively: the disorder and jumbled objects we humans must get rid of are not only superfluous and unnecessary objects, but also superfluous and unnecessary activities, pursuits and behaviours. Accordingly, the entire clause "to shed the clutter and the speed" is a non-standard collocation constituting semantic deviation (Short 1996, 43; see Sect. 2.2.1).

This interpretation is confirmed by three additional constituents in Text 5. The first is the non-finite clause "to differentiate between the essential and non-essential" (Sentence 1). The noun phrase "the essential" denotes the indispensable and fundamental elements and objects we humans require to live well; therefore, the noun phrase "[the] non-essential" indicates elements and objects which are not so—that is to say, the figurative and non-figurative clutter we have been forced to shed. The second constituent confirming the superfluous nature of clutter is the apposition "simplicity (shedding the clutter)" (Sentence 4). The noun "simplicity", viz. plainness, absence of luxury and of unnecessary objects, is glossed by the non-finite clause "shedding the clutter". The third constituent about the nature of clutter is the noun phrase "an ethical and ecological obligation" (Sentence 4). An object complement, it applies the attribute "an ethical and ecological obligation" to the direct object "simplicity (shedding the clutter)". As a result, clutter is delineated as something "non-essential", and shedding it as "simplicity" and "an ethical and ecological obligation". These linguistic data and the emphasis on the concept of shedding the clutter relay that this is a primary principle and a priority in the value-system of Text 5.

"Simplicity" explained as "shedding the clutter" is not the only gloss of a key idea offered in Text 5. Indeed, the text abounds in cases of repetition (Leech and Short 2007, 39–41; Mukařovský 1977; see Sect. 2.2.1) of syntactic structures at several levels, in nine two-part consecutive structures with the same or very similar meanings, and with the latter explaining the former. This is an instance of the foregrounding practice called over-specification by Emmott and Alexander (2016, 290). In Text 5, this stylistic technique occurs in seven sentences out of nine, at phrase level and at clause level, as shown in Table 7.1 (for the sake of completeness, the table also comprises the gloss of the noun phrase "simplicity").

Some of these nine consecutive syntactic structures and their constituents deserve further study. The finite clause "who we are" is glossed by another finite clause, "where we are" (Sentence 2). Both clauses refer to our identity as humans, the former in more general terms ("who"), the latter in geographical terms ("where"). For additional prominence and foregrounding, the two finite clauses are followed by a third finite clause, "we are members of one Earth Family", which features the repetition of the subject-predicator string "we are". These stylistic mechanisms communicate that, as far as ecological issues are concerned, the only identity that matters is our human identity as organisms living on the planet. Moreover, the noun phrase "Earth Family" is composed of two common nouns which, as such, should be written in lower-case letters; their being capitalised, consequently, constitutes an example of graphological deviation (Short 1996, 54; see Sect. 2.2.1). By

Table 7.1 List of glosses in Text 5

Sentence	Consecutive syntactic structures	First constituent	Second constituent or gloss
2	Finite clause + finite clause	who we are	where we are
2	Noun phrase + noun phrase	a common home	our beautiful Earth
3	Elliptical clause + elliptical clause	not separate from it [Nature]	nor superior to it [Nature]
4	Noun phrase + non-finite clause	simplicity	shedding the clutter
5	Non-finite clause + non-finite clause	taking simply what we need to sustain our life	living within ecological and planetary boundaries
6	Noun phrase + noun phrase	their share of ecological space	the right to participate in the living processes which ensure food and water for all
7	Finite clause + finite clause	we overstep the limits of our rightful share	disrupt planetary boundaries, ecological limits
7	Noun phrase + noun phrase	planetary boundaries	ecological limits
9	Noun phrase + three coordinated noun phrases	a deepening crises [sic] of food and water	poverty, hunger and starvation

imitating the standard spelling conventions for family names (e.g. "the Smith Family"), this deviation signals that all of us humans living on the planet are related by blood and belong to the same species.

This notion is reinforced by the noun phrase "a common home", glossed by the noun phrase "our beautiful Earth" (Sentence 2). While the graphological deviation describes our human species as a family, the Earth-as-home gloss attributes a shared family base to us (in Text 5, the noun "Earth" is capitalised in two key phrases; as a result, for the sake of consistency, it is also capitalised in this Sect. 7.2). Some of the Oxford English Dictionary senses of the entry "home" can also broadly denote the planet, especially those under the major semantic division "I. The place where a person or animal dwells"; for instance, "The place where one lives or was brought up, with reference to the feelings of belonging, comfort, etc., associated with it" (sense 2.b), and "A person's own country or native land. Also: the country of one's ancestors" (sense 5). Nevertheless, some other senses of "home" under the same semantic division define it as "a private house or residence considered merely as a building" (sense 2.a), and even as "The furniture or contents of a house" (sense 2.e). Given this varied evidence and such senses as 2.a and 2.e, it is safe to hypothesise that the Earth-as-home gloss may be regarded as literally false by several, if not many, addressees of Text 5, in particular by those who do not have an ecological mindset. Hence, to these addressees, the gloss may appear to be a case of semantic deviation (Short 1996, 43; see Sect. 2.2.1); this triggers defamiliarisation (Emmott and Alexander 2016, 291; Gavins 2014; Šklovsky 2012; see Sect. 2.2.1) and compels them, and us all, to think of the planet in an unusual and refreshing way, i.e. as a home in senses 2.b and 5.

In Sentence 6, the noun phrase "their share of ecological space" is glossed by the noun phrase "the right to participate in the living processes which ensure food and water for all". Inside these phrases, the noun phrase "ecological space" is, in turn, explained by another noun phrase, "the living processes which ensure food and water for all". Sect. 3.1 investigated the definitions of the ecology and environmental studies terms "natural capital", "natural income", "natural services", and Sect. 5.1 scrutinised that of the term "ecosystem services"; all of them were assessed against the ecosophy of this research and found to be destructive. The noun phrase "the living processes which ensure food and water for all" can be a beneficial alternative to them, especially to "natural services" and "ecosystem services". Unfortunately, it is not flawless: with its post-modifier realised by a relative clause, it is too long and complex for a technical term. However, its emphasis on food and water and on the fact that they must be secured to everyone may inspire new terms which are both technical and beneficial.

A fair number of the constituents of the consecutive syntactic structures can also be explored as examples of the stylistic strategy of point of view, specifically ideological point of view (Simpson 1993, 11–19, 27–39, 96–101; summarised in Neary 2014, 177–178; see Sect. 2.2.2). This is indicated in Text 5 by fifteen words, phrases and clauses which are positively value-laden, or which acquire a positively value-laden connotation in the ecological discourse of the text:

> [1] to shed the clutter and the speed; to differentiate between the essential and non-essential; [2] the crisis; Earth Family; share a common home; our beautiful Earth; [4] simplicity (shedding the clutter); [5] vital; [6] their share of ecological space; to participate in the living processes which ensure food and water for all; [7] our rightful share; the integrity of species; [8] the web of life; [9] justice; their rightful share

Some of these fifteen evaluative expressions require further analysis. The noun phrase "the crisis" (Sentence 2) refers to "The Corona lockdown" (Sentence 1) and to the 2019 COVID-19 pandemic outbreak; the noun phrase was categorised as positive, because the lockdown and the pandemic give us humans the golden opportunity to change our way of life and to live sustainably. Text 5 underlines the idea of sharing and presents it as an ecological tenet: all humans, animals and plants "share a common home" (Sentence 2), require "their share of ecological space" (Sentence 6) and are entitled "to participate in the living processes which ensure food and water for all" (Sentence 6) (see also "our rightful share" in Sentence 7, and "their rightful share" in Sentence 9) (for the idea of sharing in a beneficial ecological text, see also Text 4 and Sect. 6.2).

These positively value-laden expressions are present throughout the entire Text 5. On the contrary, six negatively value-laden words and phrases can only be found at the end of the text, in Sentences 7 and 9:

> [7] overstep; disrupt; [9] robbing; deprives; a deepening crises [sic] of food and water; poverty, hunger and starvation

After depicting how we humans must live during and after the pandemic, Text 5 portrays how we have been living. The complements of the lexical verbs "overstep", "disrupt" (Sentence 7), "robbing" and "deprives" (Sentence 9) consist of the positive evaluative phrases "the limits of our rightful share", "the integrity of species" (Sentence

7), "of their rightful share" and "of their basic needs" (Sentence 9); accordingly, when these complements are taken into consideration, our already unfair human actions of overstepping, disrupting, robbing and depriving turn out to be even more unfair. Furthermore, all these actions are not presented as hypothetical propositions, in the form "If we extract/extracted more from the Earth, then we will/would overstep the limits of our rightful share"; they are given in the present tense, as being factual at the present moment and possibly in the future, if we humans do not "shed the clutter and the speed" (Sentence 1) and do not behave in a sustainable and equal way.

Text 5 unambiguously suggests that it is our human responsibility to live sustainably, to share food, water and living processes, and not to rob other beings of them through another stylistic device combining point of view and foregrounding: the repetition of first-person plural pronouns, constituting proximal deixis. The pronouns "us" (two instances), "we" (seven instances) and "our" (three instances) appear twelve total times in seven of the nine sentences composing Text 5 (Sentences 1–7), for example, in the prepositional phrase "By taking simply what we need to sustain our life" (Sentence 5). All these instances are cases of inclusive we, which convey that the addresser of the text writes it from an angle and standpoint embracing all her addressees and all the public at large; in other words, none of us humans can shun these responsibilities or is excluded from this sustainability message (Simpson 2010, 295; see Sect. 2.2.2).

In addition to the addresser's ideological point of view on sustainability, Text 5 relays her psychological point of view by means of eleven modal operators, which occur in seven of the nine sentences the text consists of (for point of view and modality, see Simpson 1993, 43–47; summarised in Neary 2014, 181–182, and in Sect. 2.2.2). Two of them, "possibility" and "realising", are present in the same noun phrase in Sentence 2 ("the possibility of realising that […]"). The modal noun "possibility" is a linguistic indicator of epistemic modality, which communicates the degree of addresser assurance that we humans will realise we all belong to one species and share one planet. The lexical verb "realising" is a *verbum sentiendi* (Simpson 2010, 295; see Sect. 2.2.2), which constitutes an operator of perception modality; as such, it has the function of expressing the addresser's perceptions as based on mental processing, here what we humans have understood and have become aware of during the Corona lockdown. The value of addresser commitment signalled by the noun "possibility" is weak; that is to say, she is anything but fully confident that we are now conscious of our identity as living organisms and of our relation with the Earth.

The importance of this ecological awakening and of these ecological questions is also suggested and, above all, strengthened by the other nine modal operators in Text 5, which appear in six of its nine sentences. All convey deontic modality, or the level of addresser commitment ascribed to the performance of certain duties and tasks. The nine linguistic indicators of deontic modality are the following nouns, adjectives and lexical verbs:

> [1] invitation; [4] ethical; obligation; [5] need; [6] need; right; [7] rightful; [9] rightful; needs

Eight of these deontic modal operators relay high degrees of addresser commitment, whereas one, "invitation" (Sentence 1), communicates a low degree. This modality pattern realises positive modal shading (for modal shading, see Simpson 1993, 50–69; summarised in Neary 2014, 183–187, and in Sect. 2.2.2): the addresser gives prominence to her point of view and to what she believes our human duties and commitments to sustainability and to ecological and social justice must be.

Two of the nine sentences composing Text 5 show no linguistic indicators of modality: they are Sentence 3 ("We are part of Nature, not separate from it nor superior to it") and Sentence 8 ("In the web of life species sustain each other"). They constitute unmodalised categorical assertions, which do not psychologically judge our human attributes and identity (Sentence 3) or the interconnectedness of species (Sentence 8) signalled in the sentences; therefore, they are characterised by neutral modal shading, or the highest possible value of addresser commitment. How do deontic modality and positive modal shading interact with absence of modality and neutral modal shading in Text 5? The text is dominated by the former, which expresses what we humans must do and how we must act. Nevertheless, our actions and behaviour are determined by the assertions indicated by the latter: we must live sustainably, since "We are part of Nature", and given that "In the web of life species sustain each other".

Text 5 also conveys its sustainability message via the stylistic practice of metaphor. The semantic deviation "to shed the clutter and the speed" (Sentence 1) (see also "shedding the clutter" in Sentence 4) is founded on a conceptual metaphor (Lakoff and Johnson 2003; see Sect. 2.2.3); as a result, it can be examined by applying Steen's (2016; see Sect. 2.2.3) three-dimensional model of metaphor, as follows:

1. At the level of expression, or metaphor in language, the conceptual metaphor is activated by a predicator, the lexical verb "to shed", and by its direct object, the coordinated noun phrase "the clutter and the speed". Moreover, as previously noted, "the clutter and the speed" are described through the noun phrase "[the] non-essential" (Sentence 1). Further information about "the clutter" is also supplied by the apposition "simplicity (shedding the clutter)" (Sentence 4).
2. At the level of conceptualisation, or metaphor in thought, the direct object "the clutter and the speed" realises the target CLUTTER AND SPEED, and its combination with the predicator "to shed" the source ENTITIES TO BE SHED: this brings about the conceptual metaphor CLUTTER AND SPEED ARE ENTITIES TO BE SHED. The description of the target as "[the] non-essential" completes the conceptual metaphor as CLUTTER AND SPEED ARE NON-ESSENTIAL ENTITIES TO BE SHED. In addition, intermingling these linguistic data with those in the apposition "simplicity (shedding the clutter)" results in a related conceptual metaphor: SIMPLICITY IS SHEDDING NON-ESSENTIAL ENTITIES.
3. At the level of communication, or metaphor in interaction, the two conceptual metaphors CLUTTER AND SPEED ARE NON-ESSENTIAL ENTITIES TO BE SHED and SIMPLICITY IS SHEDDING NON-ESSENTIAL ENTITIES are both creative metaphors with a deliberate figurative function. In fact, neither is listed in the three databases Master Metaphor List (Lakoff and Cognitive Linguistics Group 1991), Metalude (Goatly and LLE Project 2002–2005) and MetaNet Metaphor Wiki (David et al. 2016–2018).

While the two conceptual metaphors CLUTTER AND SPEED ARE NON-ESSENTIAL ENTITIES TO BE SHED and SIMPLICITY IS SHEDDING NON-ESSENTIAL ENTITIES are creative, they are conceptually associated with a metaphor from Master Metaphor List and a root analogy from Metalude; hence, the evidence from the databases can contribute to studying the evidence from Text 5. Master Metaphor List catalogues the conceptual metaphor BELIEFS ARE POSSESSIONS, its special case 3 BELIEFS ARE CLOTHES and its subcategory 1 No Longer Believing is No Longer Wearing; the subcategory is exemplified by the sentence "He shed his earlier beliefs". Accordingly, in Text 5, the target CLUTTER AND SPEED can also refer to unsustainable beliefs and belief-systems to be shed and superseded by a sustainable worldview and lifestyle.

The target SPEED can also be investigated by comparing it with the linguistic data from Metalude. The most pertinent root analogy to Text 5 is INTENSE ACTIVITY IS SPEED, which contains the two lexical items "speed" and "breakneck speed". The noun "speed" has the literal meaning of "rate of movement", and the metaphorical meaning of "rate of doing an activity"; it is illustrated by the example "He can absorb information at great speed". With regard to the noun phrase "breakneck speed", its literal meaning is "travelling very fast", and its metaphorical meaning is "happening very fast"; it is exemplified by the sentence "He completed his PhD at breakneck speed". Therefore, based on this evidence, the target SPEED suggests our human activities happening very fast, and the unsustainable rates we humans engage in them; in other words, SPEED alludes to our human propensity to complete tasks in a hurried and careless manner.

The linguistic data from the three metaphor databases can also be employed to further scrutinise the graphological deviation "Earth Family" (Sentence 2). As shown above, the graphological deviation relays that we humans are linked by blood and are members of the same species. This representation can be explored by taking into account the two MetaNet frames Family and Nuclear Family (see also the frame Nurturant Parent Family, utilised in Sect. 4.2.1 to analyse Text 2) (for frames, see Tannen 1993. In accordance with common typographical conventions, the initials of frames are Capitalised, and metaphors, targets and sources are in SMALL CAPITALS). The frame Family is a subcase of Social

Group and is deployed as a source frame in such conceptual metaphors as COMMUNITY IS A FAMILY and NATION IS A FAMILY. With regard to the frame Nuclear Family, it is a subcase of Family and makes use of the related frames Authority (parents have authority over children) and Responsibility (parents have responsibilities). However, there is no stylistic clue in Text 5 evoking the Authority frame or restricting the Responsibility frame to figurative parent figures. Conversely, by means of foregrounding and point of view techniques like the repetition of inclusive we, the text makes it clear that behaving in a sustainable and equitable way is not the responsibility of just a few senior members of the Earth Family: it is the ultimate responsibility of all of us humans.

If we humans belong to the same Earth Family, viz. the same species, we share the same family home, viz. the planet. As shown above, this is communicated by the semantic deviation "a common home, our beautiful Earth" (Sentence 2), which can also be examined in terms of conceptual metaphor:

1. At the level of expression, or metaphor in language, the conceptual metaphor is prompted by the consecutive syntactic structures first noun phrase "a common home" plus second noun phrase or gloss "our beautiful Earth"; that is to say, this is another example of apposition.
2. At the level of conceptualisation, or metaphor in thought, the second noun phrase or gloss constitutes the target OUR BEAUTIFUL EARTH or, simply, EARTH, and the first noun phrase the source A COMMON HOME: this brings about the conceptual metaphor EARTH IS A COMMON HOME. The major characteristic of the source domain mapping to the target domain is that expressed by the Oxford English Dictionary sense 2.b ("The place where one lives or was brought up, with reference to the feelings of belonging, comfort, etc., associated with it"); the features signalled by senses 2.a (a building) and 2.e (furniture) do not map to the target domain. Sense 2.b also implies that we humans generally love and show solicitude for our home and, if the need arises, do whatever it takes to protect it.
3. At the level of communication, or metaphor in interaction, this conceptual metaphor is not recorded in the three metaphor databases referenced above. Nevertheless, it does not seem to be a fully creative

metaphor, neither in general discourse nor ecological discourse, for at least two reasons. Firstly, as previously noted, the EARTH IS A COMMON HOME metaphor is closely connected with the Oxford English Dictionary senses 2.b and 5. Secondly, and more importantly, its various linguistic realisations are widespread in contemporary ecological discourse; the most renowned case is probably Greta Thunberg's "Our house is on fire" special address at the January 2019 World Economic Forum in Davos, Switzerland (Thunberg 2019a). As a result, the conceptual metaphor may not be creative but, judging from how popular it has become, it is highly effective. This is confirmed by Thunberg herself quoting and elaborating on it in other speeches, for instance, the one delivered to the European Union Parliament in April 2019 (Thunberg 2019b), and the one given at the January 2020 World Economic Forum in Davos, Switzerland (Thunberg 2020), a year after her first address there. For these reasons, the conceptual metaphor EARTH IS A COMMON HOME, together with its linguistic realisations, can be termed "marker metaphor": like a marker word, its occurrence in a text indicates that that text has ecological preoccupations and belongs to the environet.

The unmodalised categorical assertion in Sentence 8 ("In the web of life species sustain each other") includes the noun phrase "the web of life", which can also be studied as a conceptual metaphor:

1. At the level of expression, or metaphor in language, the principal constituents the noun phrase is composed of are the nominal head "web" and its postmodifier, the prepositional phrase "of life".
2. At the level of conceptualisation, or metaphor in thought, the prepositional phrase constitutes the target LIFE, and the nominal head the source WEB: this results in the conceptual metaphor LIFE IS A WEB. Two main traits of the source domain map to the target domain. The first is that a web is an articulate structure consisting of a series of interconnecting threads; this trait is indicated in Sentence 8 by the fact that "species sustain each other". The second trait of the source domain mapping to the target domain is fragility: a spider's web can

easily be destroyed, like the vital balance between all humans, animals and plants sharing the planet, depicted in Text 5.

3. At the level of communication, or metaphor in interaction, the conceptual metaphor LIFE IS A WEB is not a creative metaphor. Like the EARTH IS A COMMON HOME metaphor, it is not set down in the three metaphor databases cited above; however, contemporary ecological discourse is rich in its diverse linguistic realisations. For example, a book by Fritjof Capra (1996) is entitled *The Web of Life*: this author employs the LIFE IS A WEB metaphor to explain the characteristics of organisms, social systems and ecosystems, in opposition to such conventional paradigms as mechanism and Darwinism. Furthermore, LIFE IS A WEB and its realisations are investigated by Stibbe (2021, 67–68; see also Sect. 5.1) as founded on another conceptual metaphor, NATURE IS A WEB. This metaphor encompasses us humans within nature and represents us as a part of a larger ecological system; consequently, it is our duty to realise that all our actions affect, often adversely, the other countless living components of the system. This is why the term "marker metaphor", coined to define EARTH IS A COMMON HOME, can also be applied to the conceptual metaphor LIFE IS A WEB and to its linguistic realisations: its presence in a text, like a marker word, suggests that that text pertains to the environet and is concerned with ecological matters. Text 5 is clearly such a text.

7.2.2 Summary and Evaluation

After the ecostylistic scrutiny of Text 5, drawn from Navdanya International's website, comprising and defining the marker word "sustainability", and the analysis of the stylistic mechanisms of foregrounding, point of view and metaphor featured in the text, Text 5 and these mechanisms will be summarised and evaluated. To be more precise, this section assesses the value-system of Text 5, draws a comparison with the ecosophy of this research and its norms and demonstrates whether the text and its mechanisms are actually beneficial.

As shown in Sects. 7.1 and 7.2.1, all the definitions of the noun "sustainability" and of the adjective "sustainable" furnished by the Oxford

English Dictionary and by the ecology and environment dictionaries and textbooks are ambivalent or destructive, and partly or fully run counter to the ecosophy and its norms. Hence, two new beneficial definitions in line with the ecosophy were put forward: one combining and rewording that of "sustainable" from the Oxford English Dictionary with that of "strong sustainability" from *Ecology Basics* (Editors of Salem Press 2004, 618) (Sect. 7.1); another based on Text 5 and on its contents (Sect. 7.2.1). Underlying the latter are such concepts as ecological space to be shared among all human, animal and vegetable species, the metaphorical web of life and its interconnectedness, the interrelation between sustainability and social equality. The new definition of sustainability founded on Text 5 and on these concepts follows the ecosophy and four of its norms: no. 1: Valuing Living, no. 2: Wellbeing, no. 5: Environmental Limits and no. 6: Social Justice. Therefore, both this definition and Text 5 underpinning it are fully beneficial, and the notions they convey are to be encouraged among the general public. The ecosophy and its norms are evoked and underscored in Text 5 via the stylistic strategies of foregrounding, point of view and metaphor; in several sentences, two strategies co-occur to highlight key words, phrases and clauses in the text and to leave a lasting effect on the addressee.

Text 5 opens with the foregrounding device of semantic deviation realised by the non-standard collocation in the non-finite clause "to shed the clutter and the speed" (Sentence 1). While the abstract noun "speed" designates our human haste and carelessness, the noun "clutter" can be read both non-figuratively, as "litter, jumbled objects", and figuratively, as "superfluous and unnecessary activities". "Clutter" is also portrayed by the noun phrase "[the] non-essential" (Sentence 1), and "to shed the clutter" is described as "simplicity" by the apposition "simplicity (shedding the clutter)" (Sentence 4); in turn, this apposition is credited with the quality relayed by the noun phrase "an ethical and ecological obligation" (Sentence 4). These are beneficial stylistic practices aligning with the ecosophy and its norms no. 2: Wellbeing, no. 5: Environmental Limits and no. 6: Social Justice. Norms no. 2 and no. 6 maintain that we humans must value life and the pleasures coming from basic material and spiritual means, in order to achieve a sound physical and mental condition, social equality and sustainability. As a result, we must get rid of what Odum and Odum (2001, 3, 4) called our "unessential baggage" and "useless diversions", which are near synonyms of "the clutter and the speed" and "[the] non-essential", and antonyms of "simplicity". Norm no. 5 asserts that the unprecedented global rates of consumption and waste production must be drastically reduced and kept within environmentally sustainable boundaries.

Primary phrases and clauses communicating sustainable ideas are emphasised through the foregrounding technique of the repetition of syntactic structures, in nine two-part (in one case three-part) consecutive structures with the latter glossing the former. Among them are:

1. Finite clause + finite clause + finite clause "who we are, where we are, we are members of one Earth Family", which also contains the graphological deviation "Earth Family" (Sentence 2);
2. Noun phrase + noun phrase "a common home, our beautiful Earth" (Sentence 2);
3. Elliptical clause + elliptical clause "not separate from it [Nature] nor superior to it [Nature]" (Sentence 3);
4. Noun phrase + noun phrase "their share of ecological space and the right to participate in the living processes which ensure food and water for all" (Sentence 6).

These repetitions express our human nature as embodied living beings belonging to the same species, dwelling on the same planet and depending on its natural materials and processes for our living; in other words, they remind the addressee of Text 5 that we are merely human animals sharing the Earth and its materials and processes with other human animals and with more-than-human life. These are beneficial concepts in accordance with the ecosophy and its norms no. 1: Valuing Living and no. 6: Social Justice. As claimed in the latter, at the micro-level of the species, every single human has the same intrinsic value and is entitled to the same consideration as the other humans on the planet; as noted in the former, at the macro-level of the ecosphere, we humans have the same intrinsic value and are entitled to the same consideration as the other animal and vegetable species on the planet.

More beneficial notions are also signalled by the syntactic repetition in other two-part consecutive structures with the second explaining the first:

1. Non-finite clause + non-finite clause "taking simply what we need to sustain our life, living within ecological and planetary boundaries" (Sentence 5);
2. Finite clause + finite clause "we overstep the limits of our rightful share, disrupt planetary boundaries, ecological limits" (Sentence 7);
3. Noun phrase + noun phrase "planetary boundaries, ecological limits" (Sentence 7).

The sustainable ideas indicated and reiterated by these three repetitions agree with the ecosophy and its norm no. 5: Environmental Limits. As prescribed by the norm, the repetitions underline that we humans must not cross ecological limits, and must stop consuming excessive amounts of materials, products and energy and producing waste; instead, we must only take from the Earth the bare essentials to maintain ourselves.

In Text 5, key concepts are also given prominence by means of the stylistic mechanism of point of view. For instance, the ideological point of view on sustainability of the addresser of the text is suggested by fifteen positively value-laden expressions. Some of them are utilised in Sentences 1 and 2 to evoke that "the crisis" (Sentence 2), namely the 2019 COVID-19 pandemic outbreak and the resulting lockdown, provide us humans with a propitious occasion to live in a sustainable way (see the non-finite clause "to differentiate between the essential and non-essential" in Sentence 1). Positively value-laden expressions are also deployed in Sentences 2, 6, 7 and 9 to convey that sharing the planet's materials and wealth in an equitable way is an ecological principle (see the string "share a common home" in Sentence 2). This is in line with the ecosophy and several of its norms, particularly norm no. 6: Social Justice: it points out that we humans must develop new social and economic systems by reorganising the existing ones around sustainable values and purposes, like a fairer distribution of natural materials and global wealth.

Moreover, the ideological point of view of the addresser of Text 5 is also relayed by six negatively value-laden expressions in Sentences 7 and 9. Here, we humans are depicted as engaging in the unsustainable actions of "overstep[ping] the limits of our rightful share", "disrupt[ing] [...] the integrity of species", "robbing others of their rightful share" and "depriv[ing] them of their basic needs". Not only do these unsustainable actions recall the ecosophy and its norm no. 6: Social Justice; the fact that they are in the present tense also recalls norm no. 3: Now and the Future. In fact, this norm also states that ecological issues are a reality now; consequently, urgent action to rescue the Earth must be implemented immediately, not in the future.

According to the addresser of Text 5, who is the agent accountable for actualising norms no. 6 and no. 3, along with sustainability? The text communicates that it is us humans via the foregrounded repetition of another point of view strategy, i.e. twelve first-person plural pronouns employed inclusively in Sentences 1–7. The ecosophy incorporates norm no. 4: Care, which specifically treats our human responsibility to the other human, animal and vegetable organisms on the planet; nevertheless, such a responsibility also unequivocally emerges from all the norms, especially the responsibility of those of us from the Global North. Hence, the foregrounding and point of view device of inclusive we, as utilised in Text 5 to underscore our human accountability, is beneficial and follows the ecosophy and its norm no. 4: Care.

As well as the addresser's ideological point of view, Text 5 signals her psychological point of view through modality. A noun phrase in Sentence 2 ("the possibility of realising that no matter who we are, where we are, we are members of one Earth Family and share a common home, our beautiful Earth") includes two modal operators: the epistemic modal noun "possibility" and the perception modal lexical verb (or *verbum sentiendi*) "realising". In their context, their combination indicates that the addresser is unsure whether we humans are truly aware of our identity as living beings and of our connection with the planet. As noted above, this awareness aligns with the ecosophy and its norms no. 1: Valuing Living and no. 6: Social Justice, and its weight is suggested by the beneficial stylistic practices of syntactic repetition and ideological point of view previously examined.

Another of these beneficial practices consists of the nine linguistic indicators of deontic modality which can be found in Sentences 1–9, viz. in the entire text; eight of them evoke high levels of addresser commitment and constitute positive modal shading. By means of this modality pattern, the addresser highlights our human responsibilities and obligations to sustainability and to ecological and social equality. Accordingly, the stylistic construction of an assertive addresser, who unequivocally states what her addressee must do and how they must behave, is a beneficial technique in accordance with the ecosophy. In particular, the most relevant norms to our ecological responsibilities and social obligations are no. 2: Wellbeing and no. 6: Social Justice. With a view to realising flourishing for all humans, animals and plants, we must "shed the clutter and the speed" (Sentence 1); this would permit us to only deploy "our rightful share" (Sentence 7) of the Earth's natural materials, thus also realising social equality.

The highest possible degree of addresser commitment, which constitutes neutral modal shading, occurs in the two unmodalised categorical assertions in Sentence 3 ("We are part of Nature, not separate from it nor superior to it") and Sentence 8 ("In the web of life species sustain each other"). These assertions are beneficial and agree with the ecosophy and its norms no. 1: Valuing Living and no. 2: Wellbeing (see also the beneficial definitions of the marker words "nature" and "ecology" studied in Sects. 3.1 and 6.1 respectively). The norm indeed contends that there are no distinct or self-sufficient entities in the ecosphere: it is an interconnected and interdependent complex system comprising all human, animal and vegetable species; therefore, they all support each other in a stable system and blend harmoniously within it.

Text 5 also employs the stylistic mechanism of metaphor to convey its message about sustainability. Underlying the semantic deviation "to shed the clutter and the speed" (Sentence 1) is the creative conceptual metaphor CLUTTER AND SPEED ARE NON-ESSENTIAL ENTITIES TO BE SHED; underpinning the apposition "simplicity (shedding the clutter)" (Sentence 4) is another creative conceptual metaphor, SIMPLICITY IS SHEDDING NON-ESSENTIAL ENTITIES; the fact that they are creative is likely to increase their impact on the addressee of Text 5. As shown above, the semantic deviation and the apposition, consequently the two conceptual metaphors, are beneficial and in line with the ecosophy and its norms no. 2: Wellbeing, no. 5: Environmental Limits and no. 6: Social Justice. Comparing the two conceptual metaphors with the linguistic data from Master Metaphor List and Metalude reinforced their beneficial properties: the metaphors relay that the unsustainable, hasty and careless belief-system to be shed must be replaced with a sustainable, thoughtful and careful value-system and way of life.

Another stylistic strategy in Text 5 which can also be investigated in conceptual terms is the graphological deviation "Earth Family" (Sentence 2). As previously noted, it is incorporated in a series of syntactic repetitions following the ecosophy and its norms no. 1: Valuing Living and no. 6: Social Justice. Conceptually, it is based on the two MetaNet frames Family and Nuclear Family. The former is used as a source frame in conceptual metaphors like COMMUNITY IS A FAMILY and NATION IS A FAMILY; the latter utilises the related frames Authority and Responsibility and refers them to parent figures. In Text 5, these two frames are not deployed, however: none of us humans must have authority over our fellow humans or over nonhuman organisms, and all of us must be equally responsible for living in a sustainable and fair way. These are beneficial notions aligning with the ecosophy and, again, with its norms no. 1 and no. 6.

Two conceptual metaphors in Text 5 were termed "marker metaphors": as with marker words, the inclusion of these metaphors in a text communicates that it belongs to the environet and is concerned with ecological questions. Firstly, the semantic deviation "a common home, our beautiful Earth" (Sentence 2) is founded on the marker metaphor EARTH IS A COMMON HOME. It expresses that we humans love and care for our home and will go to any length to safeguard it if the necessity arises; as a result, we must feel the same for our metaphorical home, viz. our planet. This is a beneficial metaphor in accordance with the ecosophy and its norms no. 3: Now and the Future and no. 5: Environmental Limits. In addition, the fact that our home is common to us humans and to all animals and plants recalls norms no. 2: Wellbeing and no. 6: Social Justice.

Secondly, the noun phrase "the web of life" (Sentence 8) is based on another marker metaphor, LIFE IS A WEB. It signals that a web is an extremely fragile structure composed of a complex network of interconnecting threads; hence, all forms of life on the Earth are equally fragile and mutually connected. This is another beneficial metaphor agreeing with the ecosophy and its norms, especially norms no. 1: Valuing Living and no. 2: Wellbeing: all human lives and all more-than-human lives must be valued, as all of them depend on and are influenced by one another. As brought to light by the ecostylistic examination of Text 5 and by its comparison with the ecosophy of this research, ecological sustainability is intimately related to social sustainability; as advised by Navdanya International's environmental website, we humans must strive for both ecological sustainability and social sustainability by shedding our clutter and our speed and by remembering that the Earth is our common home.

References

Allaby, M. 2015 [1994]. *A Dictionary of Ecology*, 5th ed. Oxford: Oxford University Press.

Bailey, J., ed. 2004. *The Facts on File Dictionary of Ecology and the Environment*. New York: Market House Books.

Capra, F. 1996. *The Web of Life: A New Scientific Understanding of Living Systems*. New York: Anchor Books.

David, O., G. Lakoff et al. 2016–2018. *MetaNet*. Berkeley: University of California. https://metaphor.icsi.berkeley.edu/pub/en. Last accessed April 2022.

Dawe, N.K., and K.L. Ryan. 2003. The faulty three-legged-stool model of sustainable develpment. *Conservation Biology* 17 (5): 1458–1460.

Editors of Salem Press, eds. 2004. *Ecology Basics*. Pasadena, CA, and Hackensack, NJ: Salem Press.

Emmott, C., and M. Alexander. 2016. Defamiliarization and foregrounding: Representing experiences of change of state and perception in neurological illness autobiographies. In *The Bloomsbury Companion to Stylistics*, ed. V. Sotirova, 289–307. London: Bloomsbury.

Gavins, J. 2014. Defamiliarisation. In *The Cambridge Handbook of Stylistics*, ed. P. Stockwell and S. Whiteley, 196–211. Cambridge: Cambridge University Press.

Goatly, A., and LLE Project. 2002–2005. *Metalude*. Hong Kong: Lingnan University. http://www.ln.edu.hk/lle/cwd/project01/web/home.html. Last accessed April 2022.

Kemp, D.D. 2002 [1998]. *The Environment Dictionary*, 2nd ed. London and New York: Routledge [electronic edition].

Lakoff, G., and Cognitive Linguistics Group. 1991. *Master Metaphor List*. Berkeley: University of California. http://araw.mede.uic.edu/~alansz/metaphor/METAPHORLIST.pdf. Last accessed April 2022.

Lakoff, G., and M. Johnson. 2003 [1980]. *Metaphors We Live By: With a New Afterword*, 2nd ed. Chicago: University of Chicago Press.

Larson, B. 2011. *Metaphors for Environmental Sustainability: Redefining our Relationship with Nature*. New Haven, CT: Yale University Press.

Leech, G., and M. Short. 2007 [1981]. *Style in Fiction: A Linguistic Introduction to English Fictional Prose*, 2nd ed. Harlow: Longman.

Miller, G.T., and S.E. Spoolman. 2019 [2007]. *Essentials of Ecology*, 8th ed. Belmont, CA: Cengage.

Mukařovský, J. 1977. On poetic language. In *The Word and Verbal Art: Selected Essays*, ed. J. Burbank and P. Steiner, 1–64. New Haven, CT and London: Yale University Press.

Myerson, G., and Y. Rydin. 1996. *The Language of Environment: A New Rhetoric*. London and New York: Routledge.

Neary, C. 2014. Stylistics, point of view and modality. In *The Routledge Handbook of Stylistics*, ed. M. Burke, 175–190. London and New York: Routledge.

Odum, H.T., and E.C. Odum. 2001. *A Prosperous Way Down: Principles and Policies*. Boulder, CO: University Press of Colorado.

Oxford English Dictionary (OED) Online. 2022. Oxford: Oxford University Press. https://www.oed.com/. Last accessed April 2022.

Short, M. 1996. *Exploring the Language of Poems, Plays and Prose*. Harlow: Longman.

Simpson, P. 1993. *Language, Ideology and Point of View*. London and New York: Routledge.

Simpson, P. 2010. Point of view. In *Language and Style: In Honour of Mick Short*, ed. D. McIntyre and B. Busse, 293–310. Basingstoke: Palgrave MacMillan.

Šklovsky, V. 2012 [1965]. Art as technique. In *Russian Formalist Criticism: Four Essays*, 2nd ed., ed. L.T. Lemon and M.J. Reis, 3–24. Lincoln and London: University of Nebraska Press (first published in Russian in 1917).

Steen, G. 2016. Metaphor: Metaphor and style through genre, with illustrations from Carol Ann Duffy's *Rapture*. In *The Bloomsbury Companion to Stylistics*, ed. V. Sotirova, 308–324. London: Bloomsbury.

Stibbe, A. 2021 [2015]. *Ecolinguistics: Language, Ecology and the Stories We Live By*, 2nd ed. London and New York: Routledge.

Tannen, D., ed. 1993. *Framing in Discourse*. Oxford: Oxford University Press.

Thunberg, G. 2019a. World Economic Forum 2019a special address. In *Inspirationbase: A Collection of Ethical Leadership Speeches for Research and Teaching*, collated by D. Ribitsch and A. Stibbe. https://www.ecoling.net/inspirationbase. Last accessed April 2022.

Thunberg, G. 2019b. Speech delivered to the European Union Parliament. In *Inspirationbase: A Collection of Ethical Leadership Speeches for Research and Teaching*, collated by D. Ribitsch and A. Stibbe. https://www.ecoling.net/inspirationbase. Last accessed April 2022.

Thunberg, G. 2020. World Economic Forum 2020 special address. In *Inspirationbase: A Collection of Ethical Leadership Speeches for Research and Teaching*, collated by D. Ribitsch and A. Stibbe. https://www.ecoling.net/inspirationbase. Last accessed April 2022.

World Commission on Environment and Development (WCED). 1987. *Report of the World Commission on Environment and Development: Our Common Future*. https://sustainabledevelopment.un.org/content/documents/5987our-common-future.pdf. Last accessed April 2022.

8

Summary and Concluding Remarks

8.1 Summing It All Up

The primary role of this concluding Chapter 8 is to summarise the contents of this book and the findings of this ecostylistic research and to test whether the main research purpose of the book was achieved in the five analytical Chapters 3–7 via the theoretical framework and methodology of ecostylistics, presented in the theoretical Chapter 2. Outlined in the introductory Chapter 1, the research purpose of this book is, firstly, to identify beneficial discourses (Stibbe 2021, 26–30; see Sect. 2.1.2) in the environet (Myerson and Rydin 1996, 7; see Sect. 1.1), specifically in the texts of the five environmental websites examined here (Sect. 1.2), and, secondly, to study the beneficial ecostylistic strategies they feature by evaluating these discourses and strategies against the ecosophy of this research and its seven norms (Naess 1995a, 8; Stibbe 2021, 11–16; see Sect. 1.3). A further, consequent purpose of this book is to concur with current ecostylistic and ecolinguistic research in disseminating a beneficial use of language and beneficial ecostylistic devices. With this twofold research purpose, this book focuses on beneficial discourses and practices, rather than on ambivalent or destructive ones (Stibbe 2021, 22–26).

D. F. Virdis, *Ecological Stylistics*,
https://doi.org/10.1007/978-3-031-10658-3_8

It therefore strives to contribute to a fundamental, yet undeveloped, research field in ecostylistics and ecolinguistics; it is actually prompted by the neighbouring discipline of positive discourse analysis, which aims to pinpoint positive discursive models and beneficial linguistic techniques (Martin 1999, 2004, 2006; see also Bartlett 2009, 2017; Hughes 2018; Stibbe 2018).

Against this background, this concluding Chapter 8 is based on:

1. The ecosophy of this research and its seven norms, presented in the introductory Chapter 1, namely 1: Valuing Living, 2: Wellbeing, 3: Now and the Future, 4: Care, 5: Environmental Limits, 6: Social Justice, and 7: Adaptation;
2. The theoretical framework and methodology of ecostylistics, outlined in the theoretical Chapter 2 and inspired by both ecolinguistics (Fill and Mühlhäusler 2001; Fill and Penz 2018; Stibbe 2012 and 2021) and stylistics (Burke 2014; Jeffries and McIntyre 2010; Sotirova 2016; Stockwell and Whiteley 2014). As noted in Chapter 2, the stylistic analytical approach applied here is that of practical stylistics (Carter 2010), and the stylistic mechanisms investigated are those of foregrounding (Emmott and Alexander 2016; Short 1996), point of view (Neary 2014; Simpson 1993) and metaphor (Lakoff and Johnson 2003; Steen 2016) (see also the Appendix [Chapter 9], with three checklists for these stylistic mechanisms);
3. The definitions of the five marker words (Myerson and Rydin 1996, 6) scrutinised in this book: nature, environment, ecosystem, ecology and sustainability. These marker words are defined in the five analytical Chapters 3–7 from the disciplinary viewpoints of lexicology (OED Online 2022) and ecology and environmental studies (see recent ecology and environment dictionaries and textbooks in their latest editions or latest reprints: Allaby 2015; Bailey 2004; Collin 2011; Editors of Salem Press 2004; Kemp 2002; Miller and Spoolman 2019; Smith and Smith 2015); the marker words and their definitions are introduced and assessed in the "Definition and Evaluation" sections of these chapters;
4. The results presented in the "Ecostylistic Analysis" and "Summary and Evaluation" sections of the five analytical Chapters 3–7,

which discuss a selection of linguistic data from the environet, i.e. five beneficial texts drawn from the five environmental websites Forestry England (https://www.forestryengland.uk/), Greenpeace International (https://www.greenpeace.org/international/), National Park Service (https://www.nps.gov/index.htm), Navdanya International (https://navdanyainternational.org/) and World Wide Fund for Nature (WWF) (https://wwf.panda.org/).

More precisely, this concluding chapter summarises the beneficial ecostylistic strategies employed in the five environmental websites explored in the five analytical Chapters 3–7. This is done in order to detect the key characteristics of the ecological discourses conveyed by the websites and by the five sequences selected and examined in this book. In addition, in accordance with the ecologically activist aims and scope of ecostylistics posited in this book, this chapter assesses critically, against the ecosophy of this research and its norms, the beneficial ecostylistic devices in the five environmental websites, along with the beneficial discourses they express.

The ecostylistic research in the five analytical Chapters 3–7 was carried out through the stylistic techniques of foregrounding, point of view and metaphor; as a result, each of the three following Sects. 8.2–8.4 is dedicated to one of these stylistic techniques. To be more precise, these three sections outline the major foregrounding, point of view and metaphor mechanisms evident in the five beneficial texts from the environet and from the five environmental websites including the five marker words. As shown in the five analytical chapters, such five beneficial texts are:

1. Text 1, which comprises the marker word "nature" and is taken from WWF's environmental website; it focuses on the idea of a nature-culture nexus (see Chapter 3);
2. Text 2, which contains the marker word "environment" and is drawn from Greenpeace's environmental website; it centres on our human relationship with the environment (see Chapter 4);
3. Text 3, which incorporates the marker word "ecosystem" and is taken from Forestry England's environmental website; it revolves around beavers and their ecosystemic skills (see Chapter 5);

4. Text 4, which includes the marker word "ecology" and is taken from National Park Service's environmental website; it concentrates on Traditional Ecological Knowledge, or TEK, and its interaction with the academic area of ecology (see Chapter 6);
5. Text 5, which comprises the marker word "sustainability" and is drawn from Navdanya's environmental website; it focuses on an ecologically and socially sustainable lifestyle (see Chapter 7).

8.2 Beneficial Foregrounding Strategies

Beneficial foregrounding strategies are utilised in the five texts studied in this book to convey beneficial value-systems centring on the ecosophy of this research. Various types of foregrounding, at different linguistic levels, are used in the five texts. In Text 1, the foregrounding device of parallelism is present twice. The first case, in Sentence 1 ("one [nature] contains items such as butterflies, the Amazon rainforest and photosynthesis, while the other [culture] contains items such as Beethoven's piano sonatas, wedding ceremonies or sushi"), relays the destructive concept of an opposition between nature and culture. Nevertheless, the second case of parallelism, in Sentences 3–10, denies this concept by communicating the beneficial notion of a nature-culture nexus. More articulate than the first, it is composed of four sentences asking questions about the nature-culture opposition ("What is […]", "What about […]" [twice], "Is there […]"), plus four sentences providing answers in favour of the nature-culture nexus. The question–answer sequence also constitutes a foregrounded instance of discoursal deviation, since it has the appearance of a conversation between an unlearned participant asking questions, and a learned participant answering them. This discoursal deviation demonstrates the existence of the nature-culture nexus, which is a beneficial idea aligning with the ecosophy and its norms no. 1: Valuing Living and no. 4: Care.

In Text 1, a further question on the nature-culture nexus in the concluding Sentence 11 ("So would it make more sense […]") realises foregrounded internal deviation from the question–answer parallelism in Sentences 3–10. The answer to this question is not offered by Text 1, but the addressee is explicitly required to supply it; in traditional terms, this is a rhetorical question. Another rhetorical question appears in Text 2. The interrogative Sentence 5 "Who can blame her?" literally asks the addressee who is enabled to criticise Mother Earth for being shrewish to her metaphorical human children. In fact, the sentence implies that she is not to blame: we humans are, due to our unecological behaviour. This is the only interrogative sentence in Text 2; consequently, it constitutes an example of internal deviation from the declarative sentence norm. Foregrounding interacts with point of view in these two rhetorical questions. Most of the sentences in Text 1 and in Text 2 express their addressers' point of view; however, by means of the interrogative syntactic structures and of the modal operators "would", "make more sense" and "can", the rhetorical questions directly encourage the addressee to signal their own point of view. In both Text 1 and Text 2, the rhetorical questions openly engage the addressee in the texts, so that they can play an active role in the texts and in their meaning-making processes. In the context of Text 1, the rhetorical question evokes a message which recalls the ecosophy and its norms no. 1: Valuing Living and no. 4: Care; in the context of Text 2, it evokes a message recalling norm no. 5: Environmental Limits.

In Text 2, foregrounding is also realised by the semantic deviation in the noun phrase "Mother Earth" (Sentences 2–5, 10 and 18) and by those in two three-part lists of adjectives and adjective phrases characterising the planet: "weary, disheveled and drained" (Sentence 3), and "cranky, frustrated and tired of humans" (Sentence 4). Not only do these foregrounding practices personify the planet as a woman, but they also unexpectedly personify it as an easily-angered mother; underpinning this personification is the conventionalised and lexicalised metaphor THE EARTH IS A MOTHER, which will be investigated in Sect. 8.4. While the semantic deviation "Mother Earth" activates the metaphor, the two three-part lists of her semantically deviant features turn it into the creative and deliberate, and beneficial, metaphor THE EARTH IS A SHREWISH MOTHER. This radical reversal of a common conceptualisation creates defamiliarisation and makes the new metaphor highly salient and effective. These foregrounding techniques indicate beneficial considerations in accordance with the ecosophy and its norms no. 3: Now and the Future and no. 5: Environmental Limits.

Foregrounded semantic deviation can also be found in Text 3 and in Text 5. In Text 3, beavers are described via two semantic deviations: that in the apposition "one of nature's greatest engineers, the beaver" (Sentence 4), and that in the noun phrase "ecosystem engineers" (Sentences 12 and 18). Both are semantically deviant, as the noun "engineers" generally denotes a human, not an animal, and as they join together the entities of nature and ecosystem, epitomised by the beaver, and of culture, epitomised by engineering. These foregrounding mechanisms are deployed to state that beavers, viz. nonhuman animals, as well as human animals, can design and build such complex objects as dams through their physical and mental work. This message is beneficial and agrees with the ecosophy and its norm no. 1: Valuing Living. In Text 5, the foregrounding strategy of semantic deviation in the non-finite clause "to shed the clutter and the speed" (Sentence 1) gives prominence to one of the principal concepts in the text. This prominence is confirmed by the noun "clutter" glossed as "[the] non-essential" (Sentence 1), by the non-finite clause "to shed the clutter" depicted as "simplicity" (Sentence 4) and by the apposition "simplicity (shedding the clutter)" (Sentence 4) portrayed as "an ethical and ecological obligation" (Sentence 4). This is a beneficial notion in line with the ecosophy and its norms no. 2: Wellbeing, no. 5: Environmental Limits and no. 6: Social Justice. Considering the examples of semantic deviation in Text 2, Text 3 and Text 5, it can be safely hypothesised that the function of this foregrounding device in these texts is to surprise the addressee with unpredictable linguistic collocations, hence with memorable beneficial images and considerations.

In the five beneficial texts scrutinised in this book, the foregrounding practice of repetition is employed to describe the main traits of the entities and ideas discussed in the texts and to emphasise and clarify their primary ecological topics and issues. Firstly, repetition occurs at word level. This is mostly the case in Text 2: here foregrounding and point of view combine in eleven occurrences of the inclusive personal pronoun "we" (Sentences 1 and 6–14), and in six instances of the deontic modal auxiliary verb "must" alternating with four occurrences of the deontic modal lexical verb "need" (Sentences 1 and 10b-17). Among the stylistic techniques in Text 2, these are the ones which most shape it.

This is because many instances of these words are utilised to compile a detailed list of matter-of-fact measures we humans are expected to take with a view to replenishing the environment, like "We need to decarbonise economies and shift to renewable energy systems to mitigate global warming" (Sentence 14). By means of the combination of these foregrounding and point of view mechanisms, the addresser of Text 2 suggests the entire ecosophy with all its norms, from no. 1: Valuing Living to no. 7: Adaptation. The foregrounded repetition of a linguistic indicator of point of view also typifies Text 5. The inclusive personal pronoun "we" can be found twelve times in Sentences 1–7, for example, in the finite clause "we leave vital ecological space for other beings" (Sentence 5). These occurrences demonstrate that we humans, particularly those of us from the Global North, are the agents responsible for complying with the principles of sustainability and with the norms of the ecosophy, especially norm no. 4: Care. Reminding all of us of our ecological and social accountability is a beneficial foregrounding and point of view practice.

Secondly, repetition is present at phrase level and at clause level. In Text 3, the repetition of syntactic structures at various levels appears in eleven two-part coordinated structures in Sentences 5–23, such as the noun phrases "Their [beavers'] thick, waterproof fur and broad, leathery tail" (Sentence 6). The role of this foregrounding strategy is not only to accurately depict beavers from a naturalistic perspective, but also to praise them, their bodies and their accomplishments. This specific information activates the addressee's curiosity in animal and vegetable species they may not know in general, and in beavers in particular. This is a beneficial message following the ecosophy and its norm no. 1: Valuing Living. In Text 5, the repetition of syntactic structures can be found in nine two-part coordinated structures in Sentences 2–9, with the second glossing the first, like the elliptical clauses "not separate from it [Nature] nor superior to it [Nature]" (Sentence 3). These coordinated structures convey that we humans are living beings depending on the planet and sharing its natural materials with other human animals, nonhuman animals and plants; accordingly, we are only allowed to deploy the bare necessities to keep ourselves alive and live reasonably well. These glosses are beneficial and align with the ecosophy and its norms no. 1: Valuing Living, no. 2: Wellbeing, no. 5: Environmental Limits and no. 6: Social Justice.

The foregrounding device of the repetition of syntactic structures at various levels, which characterises Text 3 and Text 5, also typifies Text 4. It occurs in four two-part coordinated structures in Sentences 3, 6 (twice) and 7, such as the noun phrases "observation and description of the empirical world" (Sentence 7). These repetitions gloss and underline beneficial facts about TEK and Indigenous culture and benefit an addressee who may not be familiar with ecology or TEK. In Text 4, a similar function to repetition is performed by the foregrounding practice of over-specification. This is present in four compressed noun phrases featuring complex premodification and postmodification in Sentences 1, 2, 4 and 5, like "a broad knowledge base of complex ecological systems in their own localities" (Sentence 4). A quality of scientific and professional discourse, compressed noun phrases relay several additional beneficial facts about TEK, Indigenous peoples and their close connection with their ecosystems in relatively few words. These facts are beneficial and in accordance with the ecosophy and its norms no. 1: Valuing Living, no. 2: Wellbeing, no. 6: Social Justice and no. 7: Adaptation. Most of the beneficial considerations communicated via foregrounding techniques are reinforced by point of view techniques, as shown in the section below.

8.3 Beneficial Point of View Strategies

The ecosophy of this research and the beneficial value-systems revolving around it are expressed by beneficial point of view strategies in the five texts explored in this book. Two types of point of view are mainly employed: psychological point of view and ideological point of view. Firstly, psychological point of view is signalled by epistemic and perception modality as follows:

1. WWF's Text 1 (with the marker word "nature"): twelve modal operators of epistemic modality (Sentences 1–11, namely the entire text), plus one modal operator of perception modality (the string "make more sense" in Sentence 11), amounting to thirteen total operators;

2. Greenpeace's Text 2 (with the marker word "environment"): six modal operators of epistemic modality (Sentences 2-10a and 18), plus one modal operator of perception modality (the adverb "undoubtedly" in Sentence 19), amounting to seven total operators;
3. Forestry England's Text 3 (with the marker word "ecosystem"): four modal operators of epistemic modality (Sentences 18–21), plus three words of estrangement indicating perception modality (Sentences 2 and 3), amounting to seven total operators.

   In Sentences 1-4, the three words of estrangement intermingle with spatial point of view to produce the stylistic effect of psychological sequencing. Furthermore, the epistemic modal operator "can" is incorporated in Sentence 20 ("Their [beavers'] return can benefit people too"): the meaning of the sentence is not beneficial but ambivalent, and must therefore be rephrased ("Their return can benefit all the animal and vegetable species living in their ecosystem, and even people not living there");
4. National Park Service's Text 4 (with the marker word "ecology"): two modal operators of epistemic modality (Sentence 6), plus four instances of citation (Sentences 2, 4, 6 and 7), amounting to six total operators;
5. Navdanya's Text 5 (with the marker word "sustainability"): one modal operator of epistemic modality (the noun "possibility" in Sentence 2), plus one modal operator of perception modality (the modal lexical verb, or *verbum sentiendi*, "realising" in Sentence 2), amounting to two total operators.

   Given this epistemic and perception modality configuration, Text 5 characterises its addresser as the one most convinced of the accuracy of the ideas she advocates and of the sustainability and social issues she deals with in the text.

Nearly all these modal operators of epistemic and perception modality in the five beneficial texts suggest medium to high degrees of modality; as such, these modal operators do not provide the texts with negative modal shading. That is to say, they do not convey the uncertainty of the addressers of the texts as to the factuality of the concepts they portray and of the ecological questions they raise in the texts.

Secondly, the five beneficial texts relay lack of psychological point of view in the following way:

1. Text 1: four unmodalised categorical assertions in Sentences 3, 5, 7 and 10, out of eleven total sentences;
2. Text 2: four unmodalised categorical assertions in Sentences 3, 4, 7 and 8, out of nineteen total sentences;
3. Text 3: sixteen unmodalised categorical assertions in Sentences 4–17 and 22–23, out of twenty-three total sentences. Owing to this modality evidence, Text 3 is the one which most expresses a veracious, rigorous and thorough description of its subject matter not psychologically evaluated by its addresser;
4. Text 4: four unmodalised categorical assertions in Sentences 1, 3, 5 and 8, out of eight total sentences;
5. Text 5: two unmodalised categorical assertions in Sentences 3 and 8, out of nine total sentences.

Lack of modal operators in these sentences from the five beneficial texts signals neutral modal shading. In other words, these sentences and the linguistic data supporting the truthfulness of the ecological ideas and topics covered in the texts are not psychologically assessed by their addressers, nor should they be by their addressee.

In the five beneficial texts, these modality patterns (no negative modal shading plus neutral modal shading) make the beneficial messages they promote more prominent. Moreover, this modality configuration indicates the addressers' overall strong commitment to the accuracy of these messages; as a result, the addressers of these texts are typified as entirely sure of their ecological considerations, viz. as writers the addressee can rely on. Because of these point of view mechanisms, the addressee's opinion cannot but tally with the addressers' beneficial one, and the addressee cannot but concur with the addressers on the importance of the ecosophy and all its seven norms.

Thirdly, psychological point of view is also suggested by deontic modality in two of the five beneficial texts:

1. Text 2: six occurrences of the deontic modal auxiliary verb "must" alternating with four occurrences of the deontic modal lexical verb "need" (Sentences 1 and 10b-17), amounting to ten total operators. These occurrences also realise foregrounded repetition, as analysed in Sect. 8.2;
2. Text 5: nine modal operators of deontic modality in Sentences 1–9, i.e. in the whole text.

Almost all these modal operators of deontic modality in these two beneficial texts evoke medium to high values of modality; consequently, these high values constitute positive modal shading. This depicts the addressers of the two texts as assertive and authoritative writers energetically telling their addressee what to do and how to behave in an environmental and sustainable way. Conveying and listing our human duties and commitments to the environment and to sustainability is a beneficial point of view technique; through this technique, the addresser of Text 5 vigorously relays norms no. 2: Wellbeing and no. 6: Social Justice, and the addresser of Text 2 the entire ecosophy with all its norms.

As anticipated above, the beneficial texts also utilise ideological point of view, which is primarily communicated by evaluative language; the only exception is Text 1, where evaluative language does not feature prominently. In Text 2, the two three-part lists of negatively value-laden adjectives and adjective phrases "weary, disheveled and drained" (Sentence 3), and "cranky, frustrated and tired of humans" (Sentence 4), examined in Sect. 8.2 as examples of foregrounded semantic deviation, surprisingly portray the earth as a shrewish mother weary of her human children. The two negative evaluative coordinated adjectives "needy and greedy" (Sentence 6), which also realise foregrounded phonological deviation, convey that we humans are responsible for plunging the planet into the environmental crisis. The positively value-laden adjective and adjective phrase in the finite clause "Earth is a fine place and worth fighting for" (Sentence 19) are the concluding words in Text 2 and urge us humans to make a concerted effort to save the planet and the environment. The notions relayed by these point of view devices are beneficial and agree with the ecosophy and its norms no. 3: Now and the Future and no. 5: Environmental Limits.

In Text 3, ideological point of view is deployed in Sentences 4–22: here beavers are described as impressive animals capable of creating a healthy and functional ecosystem by means of twelve positive evaluative noun and adjective phrases, such as "a real feat of engineering in themselves" (Sentence 22), referring to their dams. These point of view practices communicate a beneficial message following the ecosophy and its norms no. 1: Valuing Living and no. 2: Wellbeing. Furthermore, in the same text ideological point of view also appears in Sentences 9 and 10: "beavers were heavily hunted for their fur and scent glands that produce a substance called castoreum. By the beginning of the 16th century the beaver became extinct from the country". In these two sentences this point of view technique plays a different role: the two negatively value-laden string "heavily hunted" and adjective "extinct" state that we humans heavily hunted beavers and caused their extinction as a species. Nevertheless, Text 3 does not argue this directly, but via the agentless passive clause "beavers were heavily hunted [...]", whose omitted agent is constituted by us humans. In these two sentences, the text is not fully in line with the ecosophy and its norms no. 2: Wellbeing and no. 4: Care. Hence, the sentences must be reworded in a more beneficial way, for instance: "we humans heavily hunted beavers for their fur and scent glands that produce a substance called castoreum. Due to our relentless hunting activities, by the beginning of the 16th century the beaver became extinct in the country".

In Text 4, ideological point of view is also signalled by positive evaluative words and phrases; a number of them are constituents in the two-part coordinated syntactic structures and in the compressed noun phrases studied in Sect. 8.2 as cases of foregrounding. Further examples of positively value-laden words are the two lexical verbs "parallels" (Sentence 7) and "inform" (Sentence 8), contending that TEK is analogous to ecology and that it inspires conservation biology, the prepositional phrases "For thousands of years" (Sentence 3) and "around the world" (Sentence 4), and the adjective phrase "time-tested" (Sentence 5), maintaining the efficacy of TEK in time and place. These are beneficial considerations aligning with the ecosophy and its norm no. 3: Now and the Future. In Sentence 6, ideological point of view is employed to blame mainstream science for not approving of TEK. Blame is linguistically realised by the following negative evaluative words: the adverb "Yet", the epistemic modal adjective "rare", the epistemic modal adverb "often", and the two lexical verbs "overlooks" and "disparages". Mainstream science ignoring and undervaluing TEK is a destructive attitude running counter to the ecosophy; accordingly, the addresser's accusation against mainstream science and her emphasis on the validity of TEK are beneficial point of view mechanisms in accordance with the ecosophy.

In Text 5, ideological point of view is also indicated by positively and negatively value-laden words, phrases and clauses. On the one hand, of the fifteen positive evaluative expressions in the text, those in Sentences 1 and 2, like the non-finite clause "to differentiate between the essential and non-essential" (Sentence 1), convey that the 2019 COVID-19 pandemic outbreak and the consequent lockdown present us humans with an excellent opportunity to live in an ecologically and socially sustainable manner. Some other positive evaluative expressions in Sentences 2, 6, 7 and 9, such as the noun phrase "their share of ecological space" (Sentence 6), suggest the ecological tenet of sharing fairly the planet's natural materials and material prosperity among all of us humans. These point of view strategies are beneficial and agree with the ecosophy and its norm no. 6: Social Justice. On the other hand, in Sentences 7 and 9 six negatively value-laden expressions portray us humans performing such unsustainable activities as "overstep[ping] the limits of our rightful share" and "depriv[ing] them [other humans] of their basic needs". Acknowledging and condemning these activities communicates a beneficial consideration in line with the ecosophy and its norms no. 3: Now and the Future and no. 6: Social Justice. The section below shows that a fair number of the beneficial messages expressed through point of view and foregrounding devices are strengthened by metaphorical devices.

8.4 Beneficial Metaphorical Strategies

Together with beneficial foregrounding and point of view strategies, and often co-occurring with them, beneficial metaphorical strategies are deployed in the five texts investigated in this book to signal beneficial value-systems concentrating on the ecosophy of this research. In this section, the metaphorical strategies in the five texts are outlined as follows: firstly, beneficial metaphors; secondly, destructive and ambivalent metaphors rewritten in the texts themselves in a beneficial way following the ecosophy; and thirdly, destructive and ambivalent metaphors I rewrote in a beneficial way following the ecosophy.

All the beneficial metaphors not requiring rephrasing can be found in Navdanya's Text 5, which includes the marker word "sustainability". As noted in Sect. 8.2, the non-finite clause "to shed the clutter and the speed" (Sentence 1) is an instance of foregrounded semantic deviation; a related concept is indicated by the apposition "simplicity (shedding the clutter)" (Sentence 4). In metaphorical terms, the non-finite clause is founded on the creative and deliberate metaphor CLUTTER AND SPEED ARE NON-ESSENTIAL ENTITIES TO BE SHED, while the apposition is based on the equally creative and deliberate metaphor SIMPLICITY IS SHEDDING NON-ESSENTIAL ENTITIES. Both metaphors suggest that we humans must divest ourselves of unsustainable mindsets and ways of life and adopt sustainable ones instead; they also mention our human tendency to rush and do things hastily and superficially. These are therefore beneficial metaphorical practices aligning with the ecosophy and its norms no. 2: Wellbeing, no. 5: Environmental Limits and no. 6: Social Justice. Their being creative and deliberate is likely to heighten the effect they produce on the addressee of Text 5; this is the case with all the creative and deliberate metaphors in the five beneficial texts.

In the same Text 5, the noun phrase "Earth Family" (Sentence 2), which constitutes foregrounded graphological deviation, is underpinned by or conceptually associated with several metaphors and frames from the metaphor database MetaNet Metaphor Wiki (David et al. 2016–2018): the two conventionalised and lexicalised metaphors COMMUNITY IS A FAMILY and NATION IS A FAMILY, the two frames Family and Nuclear Family and the two frames Authority and Responsibility. The two latter frames refer to parents having authority over their children and having responsibilities; however, Text 5 does not convey these two frames. Indeed, what the text conveys is that we humans must not have authority over one another or over the more-than-human world, and that we must all share responsibility for living in a sustainable and equitable manner. These considerations are beneficial and in accordance with the ecosophy and its norms no. 1: Valuing Living and no. 6: Social Justice.

Text 5 also comprises two conceptual metaphors which were defined as "marker metaphors": like marker words, they indicate that the text they occur in discusses ecological issues, consequently that it pertains to the environet. The first marker metaphor is EARTH IS A COMMON HOME, which underlies the apposition "a common home, our beautiful Earth" (Sentence 2), realising foregrounded semantic deviation. This conventionalised and lexicalised metaphor communicates that we humans normally do what is necessary and required to protect our home, given that we love it and give it proper attention; it follows that we must do the same for our planet, which is the metaphorical home we humans share with all animal and vegetable organisms. This is a beneficial metaphorical technique agreeing with the ecosophy and its norms no. 2: Wellbeing, no. 3: Now and the Future, no. 5: Environmental Limits and no. 6: Social Justice. The noun phrase "the web of life" (Sentence 8) is based on the second marker metaphor contained in Text 5: LIFE IS A WEB. A conventionalised and lexicalised metaphor too, it likens the interaction among all humans, animals and plants to a fragile and complex web, thereby expressing the vulnerability and interpenetration of this interaction. This metaphorical mechanism is also beneficial and in line with the ecosophy and its norms no. 1: Valuing Living and no. 2: Wellbeing.

Moving from beneficial metaphors to destructive and ambivalent metaphors reworded in the five texts in a beneficial way aligning with the ecosophy, WWF's Text 1, incorporating the marker word "nature", shows the two creative and deliberate metaphors NATURE IS A CONTAINER and CULTURE IS A CONTAINER. They are prompted by the first example of foregrounded parallelism scrutinised in Sect. 8.2, namely "one [nature] contains items such as butterflies, the Amazon rainforest and photosynthesis, while the other [culture] contains items such as Beethoven's piano sonatas, wedding ceremonies or sushi" (Sentence 1). The two conceptual metaphors signal the notion of the nature-culture opposition and are, accordingly, destructive; however, they are only employed to outline the introductory thesis in Text 1, which will be refuted in all the remaining sentences. Indeed, the creative and deliberate metaphor NATURE AND CULTURE ARE COMMUNICATING VESSELS is elicited by Sentence 2 ("Yet nature and culture often interpenetrate and overlap"), by the foregrounded question–answer parallelism in Sentences 3–10 and, above all, by the idea of the nature-culture nexus in Sentence 11. This is a beneficial metaphorical strategy in accordance with the ecosophy and its norms no. 1: Valuing Living and no. 4: Care.

As noted in Sect. 8.2, Greenpeace's Text 2, including the marker word "environment", also includes two cases of foregrounded semantic deviation: the noun phrase "Mother Earth" (Sentences 2–5, 10 and 18) and two three-part lists of unpleasant adjectives and adjective phrases typifying Mother Earth (Sentences 3 and 4). The noun phrase triggers the conventionalised and lexicalised ambivalent metaphor THE EARTH IS A MOTHER, recorded in the Oxford English Dictionary entries "mother earth" and "mother" (for the analysis of this metaphor as ambivalent, see Sect. 4.2.1). The existence of a mother entails the existence of her children, and therefore, the noun phrase also triggers another conventionalised and lexicalised metaphor, i.e. WE HUMANS ARE THE EARTH'S CHILDREN. However, by means of the two three-part lists of unfavourable properties, the conventionalised and lexicalised ambivalent metaphor THE EARTH IS A MOTHER becomes the creative and deliberate beneficial metaphor THE EARTH IS A SHREWISH MOTHER. Why does Text 2 represent Mother Earth as far from being conventionally nurturing and protective? Because we humans, by behaving in an unduly unecological manner, have understandably angered her; as a result, the conventionalised and lexicalised metaphor WE HUMANS ARE THE EARTH'S CHILDREN becomes the creative and deliberate beneficial metaphor WE HUMANS ARE THE EARTH'S DEGENERATE CHILDREN. These two beneficial metaphorical devices agree with the ecosophy and its norms no. 3: Now and the Future and no. 5: Environmental Limits.

While Text 1 and Text 2 comprise rewritings of some of their destructive and ambivalent metaphors, I rewrote some other destructive and ambivalent metaphors in these two texts and in Text 3 and Text 4. In Text 1, the noun "manipulations" is present in the noun phrase "human manipulations to those ecosystems [the Amazon rainforest]" (Sentence 10). This noun has the literal meaning of handling recorded in the

Oxford English Dictionary sense 3.a; consequently, in the context of Text 1, it denotes the link, or rather the ecological and sustainable interplay, between the Amazon nations, the Amazon rainforest and their ecosystems. This is a beneficial concept evoking a beneficial lifestyle in line with the ecosophy and its norms no. 1: Valuing Living, no. 2: Wellbeing and no. 4: Care. The noun "manipulation", however, also has a conventionalised and lexicalised metaphorical meaning of interference and tampering, as set down in the Oxford English Dictionary sense 4 and in the three metaphor databases I consulted (David et al. 2016–2018; Goatly and LLE Project 2002–2005; Lakoff and Cognitive Linguistics Group 1991). Contrary to the value-neutral literal meaning, this metaphorical meaning is destructive and may recall the addressee of Text 1 of our unecological and unsustainable interplay with the Amazon rainforest, namely of our economic exploitation and capitalisation of it.

In order to prevent the addressee from believing that the literal manipulation of the Amazon rainforest by the Amazon nations corresponds to the metaphorical manipulation of these ecosystems by us humans from the Global North, the noun phrase in Sentence 10, quoted above, must be reworded as "human interconnections with those ecosystems [the Amazon rainforest]". The new noun "interconnections" is beneficial, as it indicates the mutual and balanced connections between the human world and the more-than-human world; hence, it confirms the ecosophy and its norms no. 1: Valuing Living, no. 2: Wellbeing and no. 4: Care suggested by the Amazon peoples' way of life.

In Text 2, the two finite clauses "we need to give Mother Earth a break—we must offer her a lifeline" (Sentence 10b), especially the noun "lifeline", activate the two creative and deliberate metaphors THE EARTH IS A DROWNING WOMAN and WE HUMANS ARE THE EARTH'S RESCUERS. On the one hand, the former conceptual metaphor beneficially describes the environmental crisis as having reached critical proportions and as requiring immediate attention, thereby following the ecosophy and its norm no. 3: Now and the Future. On the other hand, this metaphor destructively depicts the planet as an object ruled and controlled by us humans, thus diverging from norms no. 1: Valuing Living and no. 4: Care. Accordingly, this is an ambivalent metaphorical practice. With regard to the latter conceptual metaphor, it portrays us humans as heroes saving a woman's life, but it does not mention two essential facts: firstly, it is our fault that the earth is drowning in the sea, viz. facing the environmental crisis, since we are going against norms no. 1: Valuing Living and no. 5: Environmental Limits; secondly, if the earth eventually drowns in the sea, viz. if the environmental crisis is not solved, we humans and all the species she hosts will perish with her, because we are all a part of her ecosystem, as asserted by norms no. 1: Valuing Living and no. 4: Care. Therefore, this metaphorical technique is destructive.

As a result, Sentence 10b must be rewritten so as to leave out its destructive attributes. The first finite clause "we need to give Mother Earth a break" flatteringly represents us humans as heroes; consequently, it must be replaced by another clause offering a less flattering representation of us. An effective option can be the finite clause "we need to stop being the Earth's degenerate children", which is founded on one of the creative and deliberate beneficial metaphors explored above and on the norms related to it. The noun "lifeline", in the second finite clause "we must offer her a lifeline", prompts the two conceptual metaphors THE EARTH IS A DROWNING WOMAN and WE HUMANS ARE THE EARTH'S RESCUERS: given that the former is ambivalent and the latter downright destructive, the clause can readily be omitted from Text 2 without altering the beneficial meaning conveyed by the text and its aligning with the ecosophy.

Forestry England's Text 3 contains the marker word "ecosystem" and also defines it via the conceptual metaphor ECOSYSTEM IS COMPLICATED EQUIPMENT, SYSTEM, PROCESS AND/OR WORK OF PUBLIC UTILITY; this metaphor is elicited by the foregrounded semantic deviation in the noun phrase "ecosystem engineers" (Sentences 12 and 18), examined in Sect. 8.2. The metaphor is semantically and conceptually associated with the destructive metaphor NATURE IS A MACHINE and with the destructive root analogies HUMAN IS MACHINE/APPLIANCE and ORGANISATION/SYSTEM IS MACHINE; hence, it is also destructive and diverges from the ecosophy and its norms no. 1: Valuing Living and no. 7: Adaptation. To be more precise, the metaphor instils respect for the human, animal and vegetable organisms in the ecosystem only because they are conceptualised as equivalent to a human-made machine or appliance; in addition, it describes the ecosystem as a non-sentient and lifeless machine we humans are not required to take care of. These destructive elements derive from the source frame COMPLICATED EQUIPMENT, SYSTEM, PROCESS AND/OR WORK OF PUBLIC UTILITY, which in turn derives from the noun "engineers" in the noun phrase "ecosystem engineers"; accordingly, the noun phrase must be reworded in a beneficial way, so that a beneficial conceptual mechanism can be triggered.

A possibility to beneficially depict the ecosystem, based on the linguistic and naturalistic evidence supplied by Text 3, can be the frame A Beaver Ecosystem is A Complicated System with Complicated Processes Including Works of Ecosystemic Utility: this is a beneficial conceptual strategy not running counter to the ecosophy and its norms no. 1: Valuing Living and no. 7: Adaptation. Another beneficial possibility can be a creative and deliberate metaphor which, as mentioned above, is likely to be more conspicuous to the addressee of the text, therefore to make a stronger impression on them. To this end, rewriting the noun phrase "ecosystem engineers" as "ecosystem carpenters" activates the creative and deliberate metaphor ECOSYSTEM IS WOODWORK. This metaphorical device is beneficial and in accordance with the ecosophy for a number of reasons: it does not cognitively connect the ecosystem with manufactured machines or potentially dangerous appliances; it replaces artificial machines and appliances with the natural material of wood; and it underlines beavers' practical work and remarkable talent with this natural material.

The last destructive metaphor from the five beneficial texts needing beneficial rephrasing appears in National Park Service's Text 4, incorporating the marker word "ecology". The foregrounded compressed noun phrase "time-tested resource management systems of long-resident peoples" (Sentence 5) includes the conventionalised and lexicalised metaphor "(natural) resource(s)", which is highly frequent in both general discourse and ecological discourse. Sect. 3.1 proved that, in the definition provided by the disciplinary discourse of ecology and environmental studies, the metaphorical noun phrase "natural resources" designates natural materials which are vital or useful to us humans, like air, water, soil and energy. This academic definition suggests that supporting our human lives is the only goal of these natural materials; other animals and plants are not taken into account, nor is the fact that these materials have worth in themselves. As a result, this is a destructive metaphorical practice going against the ecosophy and its norms no. 1: Valuing Living and no. 4: Care.

The conventionalised and lexicalised metaphorical noun phrase "(natural) resource(s)" was recognised as destructive in the ecosophy and in its norm no. 1. Valuing Living and reworded there as the non-metaphorical noun phrases "(natural) materials", "(natural) processes", "the natural world" or, in a word, "nature". In a beneficial text and discourse, the MetaNet destructive frame Natural Resource must not be utilised to denote a natural material; similarly, the MetaNet conventionalised and lexicalised destructive metaphor NATURE IS A RESOURCE must not be utilised to denote nature. These beneficial changes to the language system result in beneficial conceptual changes. The destructive noun phrase "(natural) resource(s)" underlies several MetaNet conventionalised and lexicalised, and equally destructive, metaphors, like CONSUMING RESOURCES IS EATING and CONSUMING TOO MANY RESOURCES IS EATING TOO MUCH FOOD; underpinning them is RESOURCES ARE FOOD AND DRINK. If the beneficial noun phrase "natural materials" is substituted for "(natural) resource(s)", these destructive metaphors will become the beneficial metaphors CONSUMING NATURAL MATERIALS IS EATING, CONSUMING TOO MANY NATURAL MATERIALS IS EATING TOO MUCH FOOD, and NATURAL MATERIALS ARE FOOD AND DRINK. The FOOD AND DRINK source domain now mapping to the new NATURAL MATERIALS target domain communicates the beneficial notion that these materials are fundamental not only to our human lives, but also to more-than-human lives; consequently, we humans must appreciate them and deploy them moderately and judiciously. That is to say, vital natural materials must be utilised in agreement with the ecosophy and its norms no. 1: Valuing Living and no. 4: Care.

8.5 A Language of Connection

This concluding Chapter 8 has highlighted an essential, but representative, selection of beneficial ecostylistic strategies and discourses from the environet and their frequent co-occurrence and interaction to emphasise key ecological notions and considerations. In sum, this chapter has identified beneficial alterative models to ambivalent or destructive devices and discourses; detected practices denouncing the human-induced causes of the environmental crisis and recognising possible solutions to ecological problems; and promoted a wider use and circulation of beneficial and life-supporting techniques, discourses and, as a result, ecological agendas. In particular, the chapter has shown the ecostylistic mechanisms in the five beneficial texts with the five marker words which are utilised to convey the beneficial message that we humans are a component part of nature embedded in it; as such, we are also directly affected by the environmental crisis. These strategies subvert the deep-rooted belief conceptualising nature as an external entity, and us humans as independent, even separate, from it; they inspire us to reconsider who we are and to reformulate our ecological and social priorities. A novel interaction with the more-than-human world and the other animal and vegetable organisms on the planet implies an alteration to our mainstream human ecosophies; this must be accompanied by an alteration to our language, which must become "a language of connection" (Larson 2011, 118). A beneficial use of language and of ecostylistic devices creates beneficial discourses connecting us humans with nature and with the other humans, animals and plants on the earth; such language strengthens our awareness of how all human, animal and vegetable species are inextricably interlaced and ecologically interrelated in the environments and ecosystems we all create together. In the final analysis, we do not stand apart from nature: we are a part of it.

References

Allaby, M. 2015 [1994]. *A Dictionary of Ecology*, 5th ed. Oxford: Oxford University Press.

Bailey, J., ed. 2004. *The Facts on File Dictionary of Ecology and the Environment*. New York: Market House Books.

Bartlett, T. 2009. Towards intervention in positive discourse analysis. *Applied Linguistics Methods: A Reader*, eds. C. Coffin, T. Lillis, and K. O'Halloran, 133–147, London and New York: Routledge.
Bartlett, T. 2017. Positive discourse analysis. In *The Routledge Handbook of Critical Discourse Studies*, ed. J. Flowerdew and J.E. Richardson, 133–148. London and New York: Routledge.
Burke, M., ed. 2014. *The Routledge Handbook of Stylistics*. London and New York: Routledge.
Carter, R. 2010. Methodologies for stylistic analysis. In *Language and Style: In Honour of Mick Short*, eds. D. McIntyre and B. Busse, 55–68. Basingstoke: Palgrave MacMillan.
Collin, P.H. 2011 [1985]. *Dictionary of Environment & Ecology*, 5th edition. London: Bloomsbury [electronic edition].
David, O., G. Lakoff et al. 2016–2018. *MetaNet*. Berkeley: University of California. https://metaphor.icsi.berkeley.edu/pub/en. Last accessed April 2022.
Editors of Salem Press, ed. 2004. *Ecology Basics. Pasadena, CA, and Hackensack*. NJ: Salem Press.
Emmott, C., and M. Alexander. 2016. Defamiliarization and foregrounding: Representing experiences of change of state and perception in neurological illness autobiographies. In *The Bloomsbury Companion to Stylistics*, ed. V. Sotirova, 289–307. London: Bloomsbury.
Fill, A.F., and H. Penz, eds. 2018. *The Routledge Handbook of Ecolinguistics*. London and New York: Routledge.
Fill, A.F., and P. Mühlhäusler, eds. 2001. *The Ecolinguistics Reader: Language, Ecology, and Environment*. London and New York: Continuum.
Goatly, A., and LLE Project. 2002–2005. *Metalude*. Hong Kong: Lingnan University. http://www.ln.edu.hk/lle/cwd/project01/web/home.html. Last accessed April 2022.
Hughes, J.M.F. 2018. Progressing Positive Discourse Analysis and/in Critical Discourse Studies: Reconstructing resistance through progressive discourse analysis. *Review of Communication* 18 (3): 193–211.
Jeffries, L., and D. McIntyre. 2010. *Stylistics*. Cambridge: Cambridge University Press.
Kemp, D.D. 2002 [1998]. *The Environment Dictionary*, 2nd edition. London and New York: Routledge [electronic edition].
Lakoff, G., and Cognitive Linguistics Group. 1991. *Master Metaphor List*. Berkeley: University of California. http://araw.mede.uic.edu/~alansz/metaphor/METAPHORLIST.pdf. Last accessed April 2022.

Lakoff, G., and M. Johnson. 2003 [1980]. *Metaphors We Live By: With a New Afterword*, 2nd ed. Chicago: University of Chicago Press.

Larson, B. 2011. *Metaphors for Environmental Sustainability: Redefining our Relationship with Nature*. New Haven, CT: Yale University Press.

Martin, J.R. 1999. Grace: The logogenesis of freedom. *Discourse Studies* 1 (1): 29–56.

Martin, J.R. 2004. Positive discourse analysis: Solidarity and change. *Revista Canaria De Estudios Ingleses* 49: 179–200.

Martin, J.R. 2006. Vernacular deconstruction: Undermining spin. *DELTA: Documentação de Estudos em Linguística Teórica e Aplicada* 22(1): 177–203.

Miller, G.T., and S.E. Spoolman. 2019 [2007]. *Essentials of Ecology*, 8th ed. Belmont, CA: Cengage.

Myerson, G., and Y. Rydin. 1996. *The Language of Environment: A New Rhetoric*. London and New York: Routledge.

Naess, A. 1995. The shallow and the deep, long-range ecology movement: A summary. In *The Deep Ecology Movement: An Introductory Anthology*, ed. A. Drengson and Y. Inoue, 3–9. Berkeley, CA: North Atlantic Books.

Neary, C. 2014. Stylistics, point of view and modality. In *The Routledge Handbook of Stylistics*, ed. M. Burke, 175–190. London and New York: Routledge.

Oxford English Dictionary (OED) Online. 2022. Oxford: Oxford University Press. https://www.oed.com/. Last accessed April 2022.

Short, M. 1996. *Exploring the Language of Poems, Plays and Prose*. Harlow: Longman.

Simpson, P. 1993. *Language, Ideology and Point of View*. London and New York: Routledge.

Smith, T.M., and R.L. Smith. 2015. *Elements of Ecology*, 9th ed. Harlow: Pearson.

Sotirova, V., ed. 2016. *The Bloomsbury Companion to Stylistics*. London: Bloomsbury.

Steen, G. 2016. Metaphor: Metaphor and style through genre, with illustrations from Carol Ann Duffy's *Rapture*. In *The Bloomsbury Companion to Stylistics*, ed. V. Sotirova, 308–324. London: Bloomsbury.

Stibbe, A. 2012. *Animals Erased: Discourse, Ecology, and Reconnection with the Natural World*. Middletown, CT: Wesleyan University Press.

Stibbe, A. 2018. Positive discourse analysis: Rethinking human ecological relationships. In *The Routledge Handbook of Ecolinguistics*, ed. A.F. Fill and H. Penz, 165–178. London and New York: Routledge.

Stibbe, A. 2021 [2015]. *Ecolinguistics: Language, Ecology and the Stories We Live By*, 2nd ed. London and New York: Routledge.
Stockwell, P., and S. Whiteley, eds. 2014. *The Cambridge Handbook of Stylistics*. Cambridge: Cambridge University Press.

9

Appendix: Checklists for Foregrounding, Point of View and Metaphor

These three checklists to study the stylistic strategies of foregrounding, point of view and metaphor were developed and employed to undertake the stylistic investigation of the texts from the five environmental websites presented in the analytical Chapters 3–7. The checklists are aimed at those readers who are interested in stylistics and in the stylistic scrutiny of text and discourse, but are unacquainted with this discipline and do not have a working knowledge of its analytical tools. The checklists, which are by no means exhaustive, are provided to enable those readers to become more systematic and methodical in their stylistic analysis, in line with the tradition and approach of practical stylistics (Carter 2010, 58–61; for further information and reference, see Sect. 2.2).

9.1 Checklist 1: Foregrounding

Based on Short 1996 and Emmott and Alexander 2016
(For discussion and complete reference, see Sect. 2.2.1)

D. F. Virdis, *Ecological Stylistics*,
https://doi.org/10.1007/978-3-031-10658-3_9

Search the text under examination for the following foregrounding strategies:

- Repetition, e.g. "Blow, blow, thou winter wind" (Shakespeare, *As You Like It*, II, vii, 174)
- Parallelism, e.g. "But he was wounded for our transgressions, he was bruised for our iniquities" (*Isaiah*, 53, v)
- Discoursal deviation, e.g. not starting at the beginning of a discourse, but beginning *in medias res* (as in John Donne's "The Canonization" and Edward Thomas' "Adlestrop")
- Semantic deviation, e.g. "Come, we burn daylight, ho!" (Shakespeare, *Romeo and Juliet*, I, iv, 43)
- Lexical deviation, e.g. noun "manunkind" (e.e. cummings, "pity this busy monster, manunkind") and adjective "jollyrodgered" (Dylan Thomas, *Under Milk Wood*)
- Grammatical deviation, e.g. "Then spake King Arthur to Sir Bedivere" (Alfred Lord Tennyson, "Morte d'Arthur")
- Morphological deviation, e.g. "museyroom" (James Joyce, *Finnegans Wake*); "king- / dom", written on two distinct lines (Gerard Manley Hopkins, "The Windhover"); "Bothatonce" (e.e. cummings, "she being Brand")
- Phonological and graphological deviation, e.g. words capitalised or written entirely in capitals to suggest a prominent pronunciation (as in Ted Hughes' "Crow's First Lesson"); two black pages, a marbled page, a long series of asterisks (Laurence Sterne, *Tristram Shandy*)

The following foregrounding devices should also be looked for:

- Odd punctuation
- Under-specification and over-specification
- Interesting vocabulary and grammatical choices
- Recurrent thematic motifs
- Unusual narratological presentation
- Unusual textual layouts

9.2 Checklist 2: Point of View

Based on Simpson 1993, 2010 and Neary 2014
(For discussion and complete reference, see Sect. 2.2.2)

Search the text under investigation for the following point of view practices:

- Spatial point of view: spatial deixis, locative expressions and other techniques indicating the location or visual angle an addresser writes a text from
- Temporal point of view: temporal deixis, fast or slow pace, with or without breaks, flashbacks (analepsis), flashforwards (prolepsis), narrative gaps, instances of disnarration and counterfactuality
- Psychological point of view: *verba sentiendi*, speech and thought presentation, modality and modal shading (see below) conveying an addresser's perceptions, awareness and understanding
- Ideological point of view: language choices or stylistic and lexical selections relaying the addresser's value-, belief- and knowledge-systems

Deontic modality

- Modal auxiliary verbs

  e.g. "We *must* be a part of nature"; "We *should* be a part of nature"; "We *may* be a part of nature"
- Adjective and participle patterns: "be + deontic adjective + that"; "be + deontic participle + to"

  e.g. "It is *essential* that we be a part of nature"; "We are *allowed* to be a part of nature"

Boulomaic modality

- Modal lexical verbs

  e.g. "We *hope* to be a part of nature"; "We *wish* we were a part of nature"; "We do not *regret* being a part of nature"

- Modal adverbs
  e.g. "*Hopefully*, we are a part of nature"
- Adjective and participle patterns: "be + boulomaic adjective + that"; "be + boulomaic participle + to"
  e.g. "It is *good* that we are a part of nature"; "It is *hoped* that we are a part of nature"

Epistemic modality

- Modal auxiliary verbs
  e.g. "We *could* be a part of nature"
- Modal lexical verbs
  e.g. "We *suppose* we are a part of nature"
- Modal adverbs
  e.g. "*Maybe* / *Possibly* / *Perhaps* / *Probably* / *Certainly*, we are a part of nature"
- Adjective patterns: "be + epistemic adjective + that"; "be + epistemic adjective + to"
  e.g. "It is *certain* that we are a part of nature"; "We are *certain* to be a part of nature"

Perception modality

- Modal adverbs
  e.g. "*Obviously*, we are a part of nature"
- Adjective pattern "be + perception adjective + that"
  e.g. "It is *obvious* that we are a part of nature"

Modal shading

- Positive modal shading: higher degrees of deontic and boulomaic modal systems
- Negative modal shading: lower degrees of epistemic and perception modal systems
- Neutral modal shading: unmodalised categorical assertions, no modal operators

9.3 Checklist 3: Metaphor

Based on Lakoff and Johnson 2003
(For discussion and complete reference, see Section 2.2.3)

- Search the text under scrutiny for metaphorical expressions or linguistic realisations of conceptual metaphors
- Rewrite them in conceptual metaphor terms: CONCEPT A IS CONCEPT B, or TARGET IS SOURCE, e.g. PEOPLE ARE PLANTS; THE EARTH IS A MOTHER; LIFE IS A WEB

Apply Steen's (2016) three-dimensional approach to metaphor:

1. Expression, or the lexical, semantic and syntactic realisation of metaphor ("metaphor in language")
2. Conceptualisation, or the representation of the target by means of the source and their cross-domain mapping ("metaphor in thought")
3. Communication, or the deliberate or non-deliberate figurative function of metaphor ("metaphor in interaction")

Compare data with those listed in metaphor databases, for instance:

- Master Metaphor List (Lakoff and Cognitive Linguistics Group 1991) http://araw.mede.uic.edu/~alansz/metaphor/METAPHORLIST.pdf
- Metalude (Goatly and LLE Project 2002–2005) http://www.ln.edu.hk/lle/cwd/project01/web/home.html
- MetaNet Metaphor Wiki (David et al. 2016–2018) https://metaphor.icsi.berkeley.edu/pub/en

References

Carter, R. 2010. Methodologies for stylistic analysis. In *Language and Style: In Honour of Mick Short*, eds. D. McIntyre and B. Busse, 55–68. Basingstoke: Palgrave MacMillan.

David, O., G. Lakoff et al. 2016–2018. *MetaNet*. Berkeley: University of California. https://metaphor.icsi.berkeley.edu/pub/en. Last accessed April 2022.

Emmott, C., and M. Alexander. 2016. Defamiliarization and foregrounding: Representing experiences of change of state and perception in neurological illness autobiographies. In *The Bloomsbury Companion to Stylistics*, ed. V. Sotirova, 289–307. London: Bloomsbury.

Goatly, A., and LLE Project. 2002–2005. *Metalude*. Hong Kong: Lingnan University. http://www.ln.edu.hk/lle/cwd/project01/web/home.html. Last accessed April 2022.

Lakoff, G., and Cognitive Linguistics Group. 1991. *Master Metaphor List*. Berkeley: University of California. http://araw.mede.uic.edu/~alansz/metaphor/METAPHORLIST.pdf. Last accessed April 2022.

Lakoff, G., and M. Johnson. 2003 [1980]. *Metaphors We Live By: With a New Afterword*, 2nd ed. Chicago: University of Chicago Press.

Neary, C. 2014. Stylistics, point of view and modality. In *The Routledge Handbook of Stylistics*, ed. M. Burke, 175–190. London and New York: Routledge.

Short, M. 1996. *Exploring the Language of Poems, Plays and Prose*. Harlow: Longman.

Simpson, P. 1993. *Language, Ideology and Point of View*. London and New York: Routledge.

Simpson, P. 2010. Point of view. In *Language and Style: In Honour of Mick Short*, eds. D. McIntyre and B. Busse, 293–310. Basingstoke: Palgrave Macmillan.

Steen, G. 2016. Metaphor: Metaphor and style through genre, with illustrations from Carol Ann Duffy's *Rapture*. In *The Bloomsbury Companion to Stylistics*, ed. V. Sotirova, 308–324. London: Bloomsbury.

Index

D. F. Virdis, *Ecological Stylistics*,
https://doi.org/10.1007/978-3-031-10658-3

The manufacturer's authorised representative in the EU is Springer Nature Customer Service Centre GmbH, Europaplatz 3, 69115 Heidelberg, Germany. If you have any concerns regarding our products, please contact ProductSafety@springernature.com

Printed and bound by CPI Group (UK) Ltd, Croydon, CR0 4YY
30/09/2026
02225490-0007